D0306970

The International Karakoram Project

# Continents in Collision

## Keith Miller

**Foreword by Lord Hunt**

George Philip
London Melbourne Milwaukee

*Editor's Note*

Those particularly interested in the background to the current political and cultural problems in the Karakoram area may find it helpful to read Appendix I before the main body of the book.

The figures quoted in the text as the imperial equivalents of metric units are only approximate conversions.

*Acknowledgements*

I and all members of the expedition wish to record our sincere thanks to the many hundreds of people who by a variety of means gave us their support and made this Project possible.

*Illustration Acknowledgements*

The colour illustrations were supplied by Alan Colvill, Ian Davis, Ed Derbyshire, Bob Holmes, Keith Miller, Paul Nunn, Tony Riley, Ted Smith, Frances d'Souza, Jon Walton, Shane Wesley-Smith and Nigel Winser. The maps were drawn by the University of Sheffield Mechanical Engineering Department and the RGS drawing office. Copyright of all illustrations © Royal Geographical Society International Karakoram Project 1980.

British Library Cataloguing in Publication Data

Miller, Keith
Continents in Collision
  1. International Karakoram Project
  2. Mountaineering – Koram Range
  1. Title
  915.4'6      DS485.K2
  ISBN 0 540 01066 9

© 1982 Royal Geographical Society

Text set in 11/14½ pt Linotron 202 Sabon, printed and bound in Great Britain at The Pitman Press, Bath

# Contents

## Abbreviations

| | |
|---|---|
| APN | Agentstvo Pressy Novosti |
| BAS | British Antarctic Survey |
| FWO | Frontier Works Organisation |
| GSP | Geological Survey of Pakistan |
| GTS | Grand Trigonometrical Survey |
| KKH | Karakoram Highway |
| IKP | International Karakoram Project |
| KRC | Karakoram Research Cell |
| MEF | Mount Everest Foundation |
| NERC | Natural Environment Research Council |
| NGS | National Geographical Society (USA) |
| NWFP | North-West Frontier Province |
| ODA | Overseas Development Administration |
| PAEC | Pakistan Atomic Energy Commission |
| PIA | Pakistan International Airlines |
| PSF | Pakistani Science Foundation |
| RICS | Royal Institute of Chartered Surveyors |
| RGS | Royal Geographical Society |
| RS | Royal Society |
| SERC | Science and Engineering Research Council |
| SOP | Survey of Pakistan |
| SPRI | Scott Polar Research Institute |
| SSRC | Social Sciences Research Council |
| UGC | University Grants Commission |
| UNDP | United Nations Development Programme |
| UNMOGIP | United Nations Military Observer Group in India and Pakistan |
| WAPDA | Water and Power Development Authority |

*Figures*

*Illustrations*

*Between pp. 36–7*
General Zia, President of Pakistan, opening the Islamabad conference.
Rakaposhi at sunset.
K2, the second-highest mountain on earth.
A small rock-fall on the Karakoram Highway.
A typical Karakoram track.
The summit of Shishpare above the Ghulkin Glacier.

*Between pp. 52–3*
The Pasu group of peaks opposite the Batura Glacier.
Aliabad in Hunza with Rakaposhi in the distance.
Land Rover returning to base from Nagar.
The Karakoram Highway below Pasu.
Porters negotiating a disintegrating section of track.
Part of the Hunza gorge.

# Foreword

Exploration, in the traditional sense of making major discoveries about the visible land surface of our planet, can no longer be the principal purpose of an expedition. But there is still a need for scientific 'exploration' – the collection of data and the testing of theories in the field. In addition, an ever-expanding global population has given rise to two potentially conflicting needs which are both becoming more and more urgent. There is a need to make full use of all resources that the earth has to offer; but, in doing so, to exploit them as economically as possible. The other need – if it be true that 'man cannot live by bread alone' – is to conserve certain areas of our natural environment so as to preserve the habitats of tribal peoples, wild animals and plant life, both for their own sake and for the benefit of all life on earth. It is vitally important to preserve a balance between these two needs or, to put it another way, to ensure that they proceed in parallel. Seen in this perspective, man-made frontiers are irrelevant and can even be obstructive of the universal good.

In this context, the Royal Geographical Society, with a record of traditional exploration of which it has cause to be proud, perceives for itself a new role: to promote scientific exploration in areas that will advance man's knowledge of the world about him and as a catalyst for collaboration, both in fieldwork and laboratory research, between experts of any and every nation in those areas where the needs of conservation and exploitation are most urgent and may conflict.

Three years ago, the Society, at the invitation of the Sarawak government, enabled a large number of scientists from nine nations to undertake a series of related studies in an area of tropical rain forest in north-east Borneo, rich in timber, water resources and natural beauty, whose environmental qualities and natural habitats were endangered by the demands of industry. In 1980, the Society's 150th anniversary year, we invited experts in the earth sciences from Pakistan, the People's Republic of China, and Britain, to turn their attention to the very different phenomena awaiting further investigation in the Karakoram mountains; phenomena which endangered human life and yet which could also be seen as enriching and beneficial. As a great natural barrier delineating the frontiers between four nations and, in geological terms, a scene of dynamic and dramatic change, the Karakoram mountains were a fitting area in which to demonstrate the Society's role.

With the concurrence and co-operation of the Pakistan government and the collaboration of scientists in that country, as well as others from Britain and China, the fieldwork was successfully completed, under testing conditions of climate and terrain, during three months of intensive work. The success of this enterprise, and the spirit of comradeship which inspired it, are a striking demonstration of the universality of science. They are also a glowing tribute to the leadership of the expedition and of its various groups, as well as to the outstanding work of the Society's logistic and administrative staff.

For my wife and myself, it was a marvellous experience to be accepted as members of the happy and dedicated brotherhood of workers, during three unforgettable weeks.

*Lord Hunt KG*

# Preface

This book tells the story of the International Karakoram Project, 1980. Inevitably it is mainly a personal account as, regrettably, it is not possible to record in one volume the tales of all seventy-three members of the expedition. However, I have attempted to paint as wide a canvas as possible not only with respect to the activities of individuals involved in the expedition but also in order to illustrate the cultural, historical and political context within which we were working. The brief history of exploration in the Karakoram given in Appendix I shows only too clearly the reasons behind the long-standing political and cultural tensions in the area which were to cause us some problems. Against this background our philosophy was to promote international and interdisciplinary collaboration between scientists of several countries.

The expedition sought to bring science and technology to the Karakoram, and I think we succeeded. Certainly we introduced many technical innovations, helped to form a Karakoram Research Cell, and brought together diplomats, politicians, military personnel, university staff and the common people of Pakistan, all of whom assisted us in achieving our aims.

It was unfortunate that our motives were misconstrued and falsely interpreted by a few Russian, Indian and East European political writers, and I hope that our scientific colleagues in those countries did not also misunderstand our intentions. I am sure that this book and the two volumes of scientific papers that will result from our studies will correct this mistake; a mistake which, in the context of past history, is understandable and therefore forgiveable.

Finally, I hope readers enjoy this book, despite the overwhelming sadness that dominated our spirits, discussions and dreams during the period after the fatal accident to Jim Bishop, to whom this book is dedicated. This young Briton gave his life attempting to erect a triangulation beacon on a survey line that joins India and Pakistan to Russia and China. We derived strength to continue from the belief that a more praiseworthy activity could not be devised, and out of respect to his talent, friendship and memory, we completed our projects in full, to the best of our abilities.

*K.J.Miller*
*Sheffield, 1982*

# Introduction

Everyone knows that Mount Everest, 8,848m (29,028ft), is the highest mountain on earth. But ask someone which is the greatest range of mountains in the world and they are unlikely to answer correctly. Most people think that this honour goes to the Himalaya, the spectacular range that forms the northern boundary of the Indian subcontinent and includes Everest, Kangchenjunga, Makalu, Nanda Devi, Annapurna and many other mountains standing proudly above the 7,000m (23,000ft) level. Only mountaineers and explorers are likely to vote correctly for the Karakoram range, although this area in fact contains the greatest concentration of highest peaks to be found anywhere on the surface of this earth. This is because the Karakoram range is itself shielded by extensive mountain barriers, including the Himalaya on the south-east, the Kun Lun range of China on the north, the Pamirs on the west and the high, desolate plateaux of Tibet on the east. It was not until a hundred years ago that the full extent of the Karakoram was known; it still contains untraversed glaciers and untrodden summits, and maps of the area include blank patches. If Shangri-La is anywhere it is in the Karakoram, lost between the countless high ridges, with only one entrance and no exit, fed by streams from summit snows and with a population renowned for longevity.

Figures 1 and 2 on pages 1, 4 and 5 show clearly why the Karakoram has been

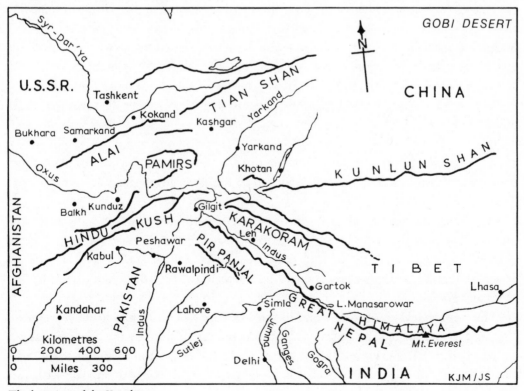

*The location of the Karakoram*

shielded for centuries, even from those who live close by. In this far corner of the earth there are eight major mountain chains, formed by a complex interaction of tectonic (continental) plate movements. This mountainous area, which marks the zone where the Asiatic and Indian continental plates meet, has the Karakoram range at its heart.

On the west is the Hindu Kush, whose peaks decrease in height as the range descends to the south-west and merges into the desert of Afghanistan. This desert is a continuation of the uniformly brown and almost limitless deserts of the Middle-East countries of Iran and Arabia, and represents a significant barrier in itself. To the north of the Hindu Kush, and separated from it by the Oxus valley, are the Pamirs, a mountain range that for many years remained more mysterious and uncharted than the Hindu Kush. Here lie the peaks of Muztagh Ata, $7,545m$ ($24,754ft$), attempted by Eric Shipton and Bill Tilman in 1947, and Mount Kongur, $7,719m$ ($25,326ft$), climbed in 1981 by four Britons: Boardman, Bonington, Rouse and Tasker. The peaks of the Pamirs rise from a high plateau frequently scoured by cold and soul-piercing Siberian winds which give meteorological support to the old Arab description of the Pamirs as the roof of the world. Here is the centre of the mountain web from which all eight ranges appear to radiate and which is the site of the continual controversy concerning the source of the Oxus. The Pamirs of Western Turkestan include the major mountains of the USSR, among them Peak Communism, $7,495m$ ($24,590ft$), and Peak Lenin, $7,134m$ ($23,405ft$).

Further north and west lie the Trans-Alai mountains shielding Bukhara, Samarkand, Kokand and Tashkent, while further north and east, providing a natural barrier between Sinkiang and Siberia, the Tian Shan runs in a north-eastern direction into the Gobi desert. South of the Tian Shan and separating the province of Sinkiang from the Tibetan plateau is the Kun Lun, famous for its gold and jade. This is the largest mountain range within China and one seldom visited by outsiders even to this day, situated as it is on the north-western edge of the Tibetan plateau.

Although the Himalaya range to the south-east of the Karakoram contains many major peaks, it is only $80km$ ($50$ miles) wide. Rising steeply from the forests of India and Nepal, it spills quickly but more gently down to the arid and high plateau of Tibet. Here rises the mighty Indus river, the greatest and only major natural resource of Pakistan, which forces its way between the Karakoram and the Himalaya through the country known as Ladakh, and then on to Baltistan (sometimes known as Little Tibet).

The eighth mountain range is the Pir Panjal, which shrouds and disguises the junction between the Himalaya and the lofty peaks of the Karakoram to the north. Although green, well forested and populated, it is only passable for a few short summer months. It is this less significant range that takes the blast of the Indian monsoon snows.

In the centre of this complex of mountains lies the Karakoram, $400km$ ($250$ miles) long and over $192km$ ($120$ miles) wide. Protected on all sides by the other mountain ranges, it forms the greatest barrier on earth to the migration of peoples. Only its lower slopes provide a little space for its many tribes. Frequently at war with one another in the past, these tribes have also not been too enthusiastic about receiving strangers. Only now is this xenophobic attitude disappearing in the major villages, as national boundaries are generally acknowledged, if not yet internationally agreed. The northern flanks of the Karakoram provide the waters of the Yarkand and Karakash rivers of China and its southern flanks the tributaries of the Indus.

Early exploration of the Karakoram range was primarily hindered by the less significant but still very difficult peaks and passes of its neighbouring ranges, each of which consists of an apparently endless succession of ridges. Each ridge, irrespective of its relative height, contains innumerable spurs that are exceedingly steep and barren, and subject to frequent landslip; these form major barriers, forcing travellers to follow an exhausting, tortuous route for many weeks up the river valleys in order to reach the mountain townships of Gilgit and Skardu, nestling below the towering peaks at an altitude of no more than approximately 1,370m (4,500ft). These two small towns are the staging posts for mountaineering and exploration of the area. It is from Gilgit that climbers set out to ascend the snow, ice and rock precipices of Nanga Parbat, 8,126m (26,660ft), 'the naked mountain' that once wiped out all but one member of a German expedition and has claimed more lives than any other mountain on earth. This mountain, more a range in its own right, is an anomalous intrusion between the Hindu Kush, the Karakoram, the Himalaya and the Pir Panjal.

From Skardu, also, a number of challenging mountains can be reached, but the most notable of them is K2. At 8,611m (28,250ft) this is the second highest mountain in the world and probably the most difficult and dangerous of them all.

It was through this extensive and ever-ascending range of barriers that explorers, traders, diplomats, missionaries and spies down the centuries have attempted to find routes to China and Russia. Here, too, there have been many attempts to establish tribal, national and international boundaries, and to defend or, depending on military strength, extend those boundaries so as to increase the size of empires. Three of the world's current major powers have in the past jockeyed for the dominant positions in this, the world's greatest natural obstacle to expansion (see Appendix I). It is hardly surprising that so little has been published on the exploration of the area, compared to the exploration of central Africa, the Americas and Polar Regions. Exploration in the Karakoram is particularly arduous – it is high and inaccessible; its internal routes can be quickly and permanently closed by snowfall until the thaws of the following spring; and its dangerous tracks are subject to frequent landslides all the year round. There was also the likelihood of a hostile reception by local peoples.

It was only in the late 1860s that sufficient details of the tortuous routes between Eastern Turkestan and India became known, and it was less than a century ago, in 1891, that the British eventually reached their most northerly boundary at the head of the Hunza Valley, north of Gilgit. It is perhaps worth noting that the area received comparatively little public attention and that colonisation did not bring commercial exploitation.

The small towns of Gilgit and Skardu were both reached in the past from the plains of the Punjab or the Vale of Kashmir, over ridges and mountain passes that

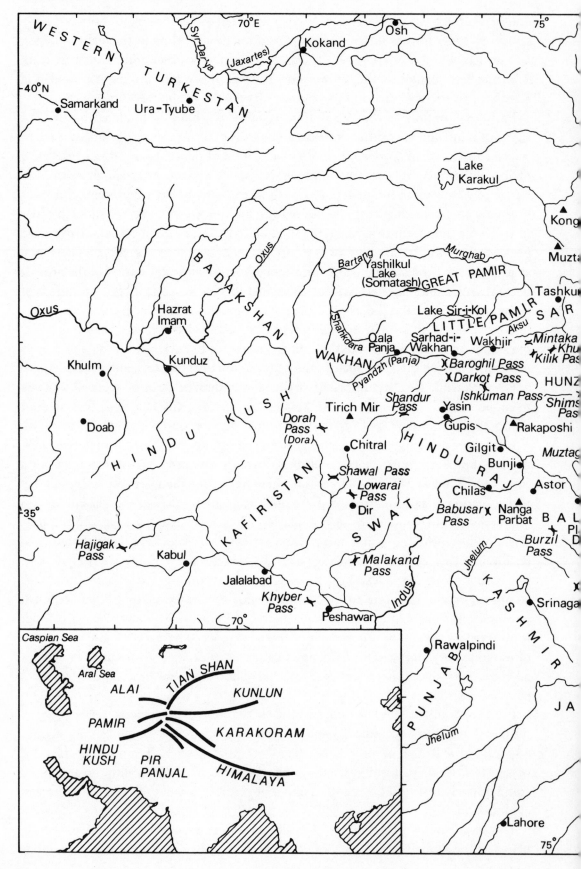

*Key locations in and around the Karakoram*

are closed except for a few weeks in the summer months. In recent years, however, it has been possible to fly into these remote mountain citadels from Rawalpindi by small aircraft whose pilots rely solely on visual navigation. If there is 50 per cent cloud cover the aircraft does not leave its base, since if visibility deteriorates further it will not be able to return the same day and could remain grounded for weeks. It is not unusual for the plane to take off but to turn for home an hour later within sight of its intended destination because of bad weather closing in from behind or ahead. Only in exceptional conditions can the aircraft follow the route over the Babusar Pass at the head of the Kaghan valley and fly under the summit shadows of Nanga Parbat, which towers 4,400*m* (14,440*ft*) above the 3,650*m* (12,000*ft*) high flight path of the aircraft.

If Gilgit and Skardu are remote, then so much more so are the base camps of the expeditions to which porters are required to carry tons of equipment for periods of up to two or three weeks. These camps are invariably placed at the foot of one of the great mountain walls that feed the Karakoram glaciers – the greatest rivers of ice outside the Polar Regions. The Siachen Glacier, which gives rise to the Nubra river, is 4.8*km* (3 miles) wide in places, over 48*km* (30 miles) long and is of unknown

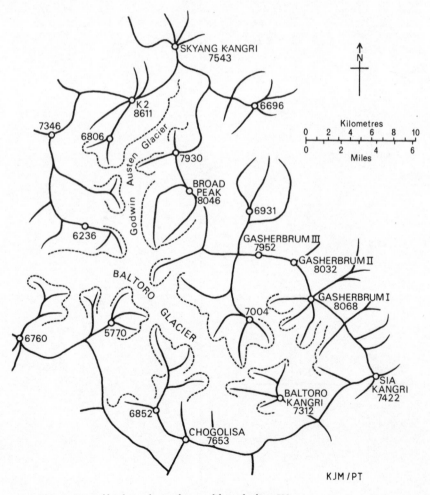

*The greatest concentration of high peaks in the world, including K2*

depth. The Baltoro Glacier is of similar proportions, and at its head rises K2, Broad Peak, the Gasherbrum peaks, and Chogolisa. In this small area alone stand ten of the world's thirty highest peaks, composed of nothing but snow, ice and rock. Many sections of these mountains frequently avalanche to feed the glaciers, which in turn gestate countless streams that eventually form destructive torrents. These mighty, unfordable glacial rivers cut through hundreds of kilometres of mountainous desert before slowing to a leisurely pace in sympathy with the far horizons and stillness of the plains of the Punjab, Baluchistan and Sind provinces of Pakistan.

It is bewildering country, and it is little wonder that the Karakoram remains largely unknown to the non-mountaineer. It was on this area that the Royal Geographical Society focused its attention in the year of its 150th anniversary. A more fitting geographical location and scientific challenge could not have been found.

# Beginnings

'We hope that Keith will lead the expedition.' So spoke the Director of the Royal Geographical Society, John Hemming, at the RGS meeting on the afternoon of 16 January, 1979.

I was taken completely by surprise. Had I known all that was to occur during the various phases of the expedition, I would have been even more cautious in my reply. I was fully aware from previous experience that all expeditions involve much time and effort before, during and after the fieldwork, and that the effort involved on the trip probably represents no more than 10 to 15 per cent of the total. As the 1980 Project was planned to be a major research expedition, the role of leader would be a greatly magnified version of any of my previous expedition commitments. Before I could take a decision of this magnitude, I had to discuss many issues with friends and colleagues, especially those who would be most affected by the outcome. I therefore told John that I was deeply honoured by his invitation, that I would give much detailed consideration to his remarks and reply within a few days.

In the previous twelve months, the Council of the Royal Geographical Society had debated several ideas for a celebratory expedition to help commemorate the 150th anniversary of the RGS, a society that has sponsored more expeditions to the farthest corners of the earth than any other single body in the world. I had previously expressed my opinion that the expedition should be interdisciplinary in a letter to John Hemming in November, 1977, immediately after a Council debate on this topic. Unknown to me, Dr Andrew Goudie of Oxford University had similar views which he had expressed in a private letter to John in 1977, at a time when I was negotiating for access to either the Karakoram or China but getting no response from the authorities in Islamabad or Beijing (Peking). Certainly all of us wished to aid the aims of the RGS, but obviously the detail of the policy of the expedition required further formulation.

Much of the long history of the RGS has been concerned with the discovery of topographical features on the earth's surface, such as the sources of rivers (for example the Nile, Niger, Amazon, Oxus), the location of the watersheds in America, Asia and Africa (which often act as tribal and national boundaries) and the determination of the position and height of mountains – in short, with filling in blanks on the map. Today, high-flying aircraft and satellites have eliminated the necessity for many of these long, demanding and frequently hazardous expeditions, and they are being superseded by multidisciplinary scientific projects. I felt that the plan we selected should reinforce and firmly establish the current trend, but also go well beyond the mere fusion of a few inter-related studies.

I hoped that we would formulate a project in which modern technological advances could be applied to geographical exploration and research. Certainly the time was right for such a venture, since the RGS now had much experience of interdisciplinary expeditions, notably to the rain forests of Borneo and the Mato Grosso of South America, the latter in collaboration with the Royal Society. Being an engineer, however, I wanted to widen the scope of such expeditions still further by injecting engineering science and its associated disciplines into projects concerned with surveying and the earth sciences.

Consequently I had proposed that we should study a geographically attractive area, list its outstanding problems and then see how recent technological developments could assist in the solution of those problems. As a mechanical engineer, I was aware of several useful developments that could be applied to subjects of a geographical nature, which would help correct the false image of geography as a qualitative, non-mathematical and hence non-scientific discipline. My ideas were not too difficult to propound, since several expeditions which I had recently completed to the glaciers of Greenland and the ice-cap of Iceland were based on this philosophy.

Thus, on 30 October 1978, the RGS agreed to examine the possibility of a large, multi-faceted expedition to the Karakoram, in which several research projects would be conducted simultaneously. The proposal was to use the Karakoram as a testing-ground for theories about continental drift, mountain building and decay, and to study the effect of ever-present hazards on the local population. The expedition would be interdisciplinary, that is, the members of each project team were to seek relationships with all the other programmes of work, rather than confine themselves to a narrow and specific outlook. Geographers, engineers, physicists and surveyors would be encouraged to learn from each other as well as to make individually significant contributions. This was in sympathy with the views of Andrew Goudie, who had proposed a geomorphological series of studies in the Karakoram to follow up the pioneering work of Kenneth Mason in the early 1900s.

As for the choice of a leader, although several members of Council had experience of the Karakoram, I thought it was a marvellous opportunity for a younger man to take the reins and thereby gain valuable experience that could be put to use by the Society during the next two decades. I had in mind a contemporary version of Gino Watkins, the young British explorer lost in Greenland's icy waters in 1932; indeed, I had already proposed the names of three young men who had been with me on various expeditions to the Arctic, all of whom were more than capable of implementing the policy we had agreed.

The method by which the RGS selects its leaders for an official expedition is known only to a few, and in the past mistakes have been made. I did not know who had been consulted before my name had been tabled; neither did I know if friends from previous expeditions had been questioned about my suitability. I was not a well-known figure in the field of exploration and had studiously avoided writing any books recounting my adventures.

First I talked to Geoffrey Sims, the Vice Chancellor of my university at Sheffield, who gave me all the encouragement I needed. Then I talked to Jim Bishop, one of the three men I had recommended as a possible leader several weeks previously. Finally I talked to John Hemming and told him of my limitations, so that he could withdraw the offer if he wished. In particular, I told him that although I would put all my efforts into advancing the policy I had suggested, I was not a leader in the usually accepted sense. Most of my previous expeditions had been relatively small, requiring minimal logistics and rapid movement. I had not previously led a national venture, although I had led British-based expeditions that had included members from other nations. I could never be a dictatorial leader and would rather discuss a point of contention before making a decision. Also, I did not wish to be regarded as a mountaineering leader, since the job of co-ordinating a team effort and placing one or two men on a high summit calls for particular skills. Admittedly, I had been on two mountaineering expeditions to the Karakoram, the first with Eric Shipton in 1957 to the Siachen and the second with a four-man expedition to K12 in 1960, but this was to be a primarily scientific expedition. The main motivation of the individual expedition members would be the acquisition of experimental data at any time of the day or night, sometimes disregarding the requirements or abilities of the team as a whole. Although I knew they would be fit, determined, sporting and chivalrous, I foresaw that they would also be demanding, as many academic prima donnas are.

I knew that on a scientific expedition in the Karakoram mountains flexible leadership would be needed to smooth out conflicting personalities; but the expedition would also need a leader who would not shirk hard decisions if and when the need arose. Moreover, because of the size of the 1980 venture, it would be essential to delegate many responsibilities to Field Directors, who would oversee the individual scientific programmes. My major role would be to look after the various diplomatic and political aspects of the expedition, to help in the selection of members, to assist in formulating research proposals, to help acquire funds and to keep a balance between conflicting interests. I had had some experience in dealing with government agencies in several countries during the course of my other ventures, and I thought that our major problems would occur before we left the UK. If these could be solved, then the only substantial problem remaining would be to ensure that sufficient work was carried out to enable us to launch future projects based on our philosophy and achievements.

I was, however, troubled by the question of my own personal fitness, an issue I kept to myself. In 1975 Jim Bishop, Chris Padfield, John Thorogood and I had made the first ever crossing, north to south, of the Staunings Alps of north-east Greenland. In three weeks, sometimes carrying 32kg (70lb) packs up steep 450m (1,500ft) snow and ice gullies, over unknown passes and glaciers, we successfully completed a 450km (280 mile) traverse; but back home in the UK my right knee started giving trouble and the specialist told me that I had 'old-man's' or 'footballer's' knee and that this needed a recovery period. Somewhat fortuitously, for the first time in eight years I did not return to the polar regions in 1979, since I had heavy commitments in Cambridge, but I was worried that I might lose my fitness while resting the knee. Perhaps selfishly, I put this doubt to the back of my mind and hoped that the experience of my previous sixteen expeditions to various parts of the world would carry me through.

Despite my acknowledged limitations, John Hemming still wished me to take the job, and so I accepted the challenge and immediately began to involve myself in the many tasks that now required action. We did not have a long time to plan the venture, but fortunately John had already made a preliminary but detailed reconnaissance in Pakistan during mid-September, 1978, and had tabled his report to the Council. I was to follow this up by visiting Islamabad after our Karakoram committee had been formed, expedition programmes designed and skeleton teams selected. In particular, I wished to follow up as soon as possible the contacts that John had so painstakingly gathered on his way back to the UK from the Mulu expedition in Sarawak.

The thirteen-member committee (see Appendix II, p. 192) was chaired by John Auden, who had spent twenty-seven years in the service of the Geological Survey of India. This lively and warm man had had much experience of the high mountains, having been six times to the Karakoram and Himalaya between 1932 and 1939, and he gave me considerable support throughout all the difficult planning phases. John Hemming and John Hunt (President of the RGS) did likewise, but they also had to consider the wider interests of the RGS. They were aware, probably more than I, that I had to balance the interests of the expedition with those of the RGS and that when difficulties arose I would be probably biased towards the expedition. I was therefore very thankful to receive their full support and their acknowledgement that I should get on with the job in my own way. Most, if not all the committee felt the same way, since they had substantial expedition experience. In consequence, only a few issues created difficulties, the major one being finance.

I have never yet been associated with an expedition that did not worry about money and the Karakoram Project was no exception. While the RGS is a most prestigious body, it does not have the funds to support an £80,000 venture. In 1978 the successful RGS expedition to Mulu was still many tens of thousands of pounds short when it left the UK for its fifteen months' work in the forests of Sarawak in Borneo. My financial policy for the 1980 Karakoram Project was to draw up research proposals which would for the first time raise substantial funds for the RGS from the various UK university research councils. This would serve two purposes. First, the Field Directors would have to propose plans in depth at an early stage which would be reviewed by their academic peers and which would be formulated in such a way as to show how their programme would interact with other disciplines. Secondly, it was clear that in an ever-worsening economic climate only the very best research programmes were now being funded. Therefore, if we received the support of the Royal Society (RS), the Science and Engineering Research Council (SERC), the Natural Environment Research Council (NERC) and the Social Sciences Research Council (SSRC), then we could hope for more funds from other sources for what would now be nationally accepted research aims. If funds could not be provided for a particular study from a research council, then we would have to accept that it did not have sufficient merit and it would have to be dropped from the plans. Six projects were eventually decided upon in the areas of geology, seismology, surveying, geomorphology, housing and natural hazards and glaciology.

Fortunately, all six scientific programmes eventually received the necessary support, although one programme which we thought would be oversubscribed only received 50 per cent of its requirements, and that not until the first days of the expedition itself. Another programme accumulated far greater debts in the field than anyone ever expected, but by that time we could just afford to keep it going to the end of the Project. The RGS made a significant financial contribution, promised indirect aid through the use of its offices and also permitted Nigel Winser and Shane Wesley-Smith, two of its administrators, to spend a good proportion of their time on Karakoram plans.

Research proposals take time to compose, and it is important that they reflect all possible modern research attitudes and are completely up to date with the state of the art. Applications for grants should therefore only be posted a few days before pre-set deadlines. Furthermore, our proposals had to be vetted by consultative panels (see Appendix II, p. 192) and built around the skills of the expedition members, some of whom still had to be selected. I was also faced with the additional difficulty that the RGS had recently launched a centenary appeal; it would have been improper to approach companies and institutions with a request for funds when they were still considering the first appeal. Accordingly, by January 1980, on the day that I arrived back from a fund-raising exercise in the USA for my own engineering research work in Sheffield, I was met by a sad committee. The expedition finances read:

*Expedition costs £75,000*
*Expedition income £16,000*

After the meeting I was politely told that this was not good enough. In reply, I said that to my knowledge no expedition was financially stable at this stage of the game, that our policy was sound, that if any single programme was not funded everyone was aware it would have to be abandoned, and finally that we had done our homework with regard to the correct procedures to gain the support of all the research councils in the UK and that their deliberations could not be hurried. I informed the committee that in the current situation I was quietly optimistic, a sentiment based on many discussions held in various parts of the UK with team members, consultants, research council secretariats and other sponsoring bodies. Although serious doubts were to remain until our crucial committee meeting on 20 March 1980, this did not prevent us from proceeding with the team selection.

The selection of Project members was based on a number of criteria. The key appointments, of course, were the Directors of the individual scientific programmes, because they, more than anyone else, would be responsible for the detailed formulation of specific research tasks and the raising of funds to carry out those tasks. The selection of Directors was also important because from them would be selected one or two Deputy Leaders who would have to take control of the entire Project should I have to withdraw for any number of possible reasons before or during the expedition. This aspect weighed heavily on my mind, since at the time I was very busy on various projects at Sheffield University.

The Directorships were settled quickly. The first Director to be appointed was Dr Andrew Goudie, who was to take control of the geomorphology programme (and was also a Deputy Leader). He had been one of the originators of the Karakoram Project and in committee was seen to be direct, clear-thinking and decisive. These qualities are often attributed to an aggressive nature, and I initially made this erroneous assumption about Andrew. However, as the time for departure drew closer, I realised my mistake and saw many other positive qualities in him, namely, an experienced and efficient field worker, a progressive and hard-working academic and finally, a young man who had a great sense of humour coupled with a deep understanding of and sympathy with his colleagues. He was one of the few members in the field who totally appreciated the diplomatic/political tightrope we had to walk and who could make rapid, sensible judgements at times of stress. Another of his attributes was that he did not interfere with problems outside his own zone of activities, but always offered sound advice if consulted.

The second appointment was Ron Ferrari, of the Engineering Department at Cambridge University. He was to manage the glaciology research. Ron and I had been Fellows together at Trinity College, and had shared two expeditions on Vatnajökull, Iceland, in 1976 and 1977.[1] Ron, the oldest member of these two expeditions, never once lost his commitment to the task in hand, his smile or his willingness to assist in every sphere of expedition life. Many misinformed members of the public consider that university staff inhabit ivory towers, protected from the hardships of the world. Ron's work on Vatnajökull, however, involved fiddling around with incredibly small electronic devices, while lying down on wet, cold ice for prolonged periods, no fun by any standards.

Ron's expertise lay in the field of radar devices and, being an academic, he was able to apply for research funding to NERC to study the depths and profiles of glaciers. Fortunately I had many good friends on the staff of the Scott Polar Research Institute (SPRI) and the British Antarctic Survey (BAS) and Ron and I had lengthy discussions with these groups in an attempt to design a collaborative programme that would eventually bring useful results to everyone. To this end, a young engineering graduate, George Musil, left Sheffield to start work at BAS on developing a radar instrument for use, initially in the Karakoram in 1980, but eventually on the Antarctic continent during 1981–2.

Sadly Ron himself had to withdraw from the Project; I attempted to dissuade him on several occasions, but eventually I had to be content for him to come out to the Karakoram for three weeks in order to see some fruits of his many months of labour. Meanwhile, however, I had the task of convincing NERC not to withdraw the research grant. This would in fact have been disastrous, since at that time it was the one and only grant we had been awarded. Ron helped by agreeing to continue acting as the academic supervisor of the research work. Fortunately, a member of the radar group, Gordon Oswald, was available to take over the Directorship, but only for the initial six weeks of the expedition. Gordon had several years' experience in the radar/electronics arena, both as a consultant and as an employee of a leading British company developing systems to sound the depth, not only of ice, but also of soil. Gordon's immediate task was to select another member to replace him as Director when he returned to his company. He selected a proved and reliable friend, Marcus Francis. The easy-going nature of both Gordon and Marcus was reflected in their willingness to accept Jim Bishop and myself in their team without complaint or discussion. Gordon and Marcus also had the ability to go native, if only in their attire, and while they were sometimes difficult to recognise, this attitude epitomised their total involvement with the environment we wished to study in depth.

On being appointed as Leader of the Project, I had immediately contacted Jim Bishop and explained the embarrassing position I was in, having previously recommended *him* to the RGS as one of three possible leaders. We had been in many scrapes together and had relied on each other's skills on countless occasions in Greenland and Iceland. Jim, in a typically unselfish manner, congratulated me and offered his services in a number of spheres. At the beginning of the planning stage of the Project I talked over my plans frequently with him, and I also had many conversations with Chris Padfield, who had been a partner in our 1975 expedition to cross the Staunings Alps of East Greenland.[2] Although Chris was due to work overseas soon and was therefore unable to join us, he was enthusiastic and wished to involve a mutual friend, James Jackson, who eventually became Joint Director of the seismology programme. Jim, Chris and James wished to broaden their own horizons, even if only in planning, and suggested that it would be useful if we could include a study concerned with the local population, particularly on how they took account of the risk of earthquakes when constructing their homes. Deeply concerned with the welfare of peoples in developing countries, they wanted to set up an integrated programme that would include the study of earthquakes, house design and an awareness of natural hazards. Such a philosophy was very much in line with what we were attempting to achieve overall, and within a few weeks discussions started with Ian Davis of Oxford Polytechnic, an expert on housing in developing countries prone to natural catastrophes such as floods, earthquakes, landslides, etc. At the same time I approached two old friends, Dr Robin Spence and Professor Cliff Moughtin, both of whom had had experience in developing countries related to the application of appropriate technology (for example, the use of local, low-cost materials and the architectural analysis of simple house structures).

Fortunately, everyone concerned quickly became friends and an apparently unending stream of memoranda appeared on my desk after each meeting of this group. Ian, now appointed as a Director of what was eventually called 'The Housing and Natural Hazards Group', was a hard-working liberal democrat greatly admired by his team, but I became a little worried by the rate at which plans were being developed, some may say altered. However, the many meetings that this group had before the expedition were crucial in establishing a common approach and in ensuring that all members understood the five diverse disciplines represented :

anthropology, architecture, engineering, geology and seismology. Unfortunately, it eventually became necessary to rationalise the group's activities, which meant hard decisions about a reduction in membership. Fortunately, Ian and I were in agreement. We decided to concentrate on very specific fields of activity, and to fix two definite programmes on which to base our negotiations with the UK research councils and the Pakistan universities and government. One was engineering-orientated and was the prime concern of Robin; the other, based on community studies, was the responsibility of Ian, who was also concerned with the overall synthesis of the programme. This self-contained interdisciplinary group turned out to be a very happy team and completed a very extensive and worthwhile programme.

About this time, James Jackson had involved his departmental colleague, Dr Geoff King, in planning an earthquake study. Geoff came down to the RGS and explained in a confident talk why a programme of this kind in the Karakoram was of importance in terms of tectonic plate theory. I for one was surprised to learn that he considered the Karakoram fault, a symptom of the tectonic plate movements in the area, to be comparable to the San Andreas fault in California and that no-one was sure of its full extent or influence. Geoff and James had excellent working relationships with experts world-wide and it was conceivable that collaboration could be solicited with neighbouring countries monitoring seismic shocks in the area.

Unfortunately, Jim Bishop now had to resign, since his firm wished to assign him to a responsible position overseas. This was a great disappointment, because I had come to rely heavily on his judgement. Also, his unique experience in glaciology and surveying after three years in Antarctica, and his involvement with the Housing and Natural Hazards Group as a civil and structural engineer made him an almost impossible man to replace. As it turned out, he was able to rejoin us at the last minute.

The impossible was achieved by one of those coincidences that defy description. Jim had shared a tent for many months in Antarctica with Jon Walton, and as a direct consequence of this friendship Jim had married Jon's sister, Jane, whose twin sister, Myra, had visited Greenland to extend some of my early 1970s surveys. A Manchester University lecturer, Dr Ted Smith, had been to Greenland with me in 1973 and knew Myra and thus the entire Walton family, who can be simply described as 'expedition-orientated'. Only two days after Jon returned from Antarctica, Ted asked him if he was interested in visiting the Karakoram, since the Survey Group needed a Director and several experienced surveyors. Jon gladly and wholeheartedly took up the challenge and immediately after taking up the Directorship organised the survey team into a closely-knit group, all of whom had training and experience with modern survey instruments. He invited Ted Smith, Alan Colvill and John Allen to join the team, all of whom had previously worked with me in Greenland. Both Jon and I knew they would seize the opportunity with both hands and produce excellent work.

Finally, we asked Professor Tahirkheli of Peshawar University to become a Director and co-ordinate the work of Pakistani geologists involved in the Project.

Tahirkheli can only be described as an enthusiast. He is very hard-working, but his most endearing and overpowering quality is the intensity of his cheerfulness – his smile seems to cover his entire face. We were particularly pleased to have him as a Director, because he is a constant and well-motivated worker in a country where there is a lack of trained men able to plan and execute a prolonged field programme. He and his colleague, Qasim Jan, proved to be two Pakistani scientists of high calibre, who could make a significant contribution to our international Project. They had had many seasons working in the Karakoram and knew many of the snags we were likely to encounter. Furthermore, they were firm allies in overcoming several of the difficult issues in the negotiations between ourselves and the many authoritative bodies in Pakistan. Finally, since the geology of the Karakoram has received relatively little attention, given its size, their programme was a most welcome addition to our research and helped to achieve our aim of an integrated international series of studies.

Each Director was responsible for selecting the members of the expedition who were to work with them. The award of any future research grant, and hence the future development of an academic career, is closely linked with past successes. The Directors therefore had to choose members who would work to everyone's benefit, taking into consideration motivation, experience, research potential, temperament, availability and the ability to appreciate other disciplines and so interact with the members of other programmes. Finally, it was of paramount importance that members should be able to contribute to, and accept team decisions. Obviously, no person yet born has all these qualities, and indeed some of them are mutually exclusive. For example, a young person may have tremendous research potential judged on his or her undergraduate career, but have minimal research experience. Similarly, a person may be powerfully motivated in a very specialised field of activity, and be an internationally known and respected worker on this topic, but may not have time enough to devote to other members in the same general field, let alone time for other disciplines.

After deciding the most important criteria and their order of priority, Directors eventually selected their own members, in the process consulting their own board of consultants, other Directors, members already selected and myself. In retrospect, their judgments were based principally on the criteria of easy-going temperament, experience and work-rate. These qualities can only be gauged by intimate knowledge of each person, and so, understandably, a large number of mutual friends were selected, the argument being that it was far better to deal with a known, rather than an unknown quantity. Space alone does not permit me to recount the recruitment and history of all the characters of this adventure, since there were in total some seventy participants, but a few details of the members are given in Appendix II and elsewhere in this book.

However, all members would agree that one Field Director deserves a special mention. Before the Karakoram Project, Nigel Winser already had several expeditions to his credit, and in 1977–8 he had been Field Organiser of the RGS Mulu expedition. Early in 1979 he came to Sheffield for a weekend to talk over his views on how to plan the venture. In any team there are differences of opinion, and our expedition was to be no exception. Furthermore, I am not a willing committee man. One problem of modern society, as I see it, is that all opinions must have equal sway, all aspects have to be accommodated, and 'democracy' is related to the size of the committee voicing the opinions. The result, invariably, is compromise rather than the best solution. Fortunately, Nigel and I had identical approaches to the solution of most of our problems, namely, that actions should always be taken with the efficiency of the whole Project as the major criterion. To both of us democracy meant ensuring that everyone had some mechanism through which to give an opinion. Apart from the comradeship that easily and rapidly pulled us together in those early days, Nigel quickly exhibited his organisational skills. Far more experienced in logistics for big expeditions than myself, he set about his multifarious tasks with Herculean energy; and he was in the right place to do it, since he was employed by the RGS to help people from any walk of life to organise their expeditions to the far corners of the world. For fourteen months he gave generously of his time and energy to the Karakoram Project. Problems to do with transportation, shipping equipment – scientific, mountaineering, medical, clothing, culinary – administration, public relations; all came to his desk marked for immediate action. Reports, minutes, memoranda, brochures, applications, a vast correspondence, and finally the all-important schedules for action were speedily and efficiently despatched. The expedition owes him a great debt of gratitude for his efforts, which some members were not fully aware of until the end of the Project.

Nigel was more than ably assisted by Shane Wesley-Smith, who matched his commitment, charm and skill during each hour of the day over her entire three-year involvement with the Project. Shane, working in the same RGS office as Nigel, would accommodate every request, however difficult and inconvenient, without a murmur, although we have subsequently joked over three incidents when she found it necessary to put Professors, Senior Military Officers and Diplomats in their place when they interfered with the running of the Project. Her experience shone like a bright light during the last few frantic days before our departure when, surrounded by chaos and sleep-starved expedition members, she shepherded us all to Heathrow Airport with tons of baggage and saw us safely aboard the British Airways plane to Pakistan.

Nigel and Shane are, without doubt, the finest, most long-suffering, most hard-working administrative team I have ever known. Churchill's cry 'Action this day' was their motto. It was sad that on the last few days of the Project they contracted hepatitis and were temporarily forced to drastically change their lifestyle on return to the UK in a debilitated state.

# Reconnaissance

The turbulent history of exploration in the Karakoram (see Appendix I) clearly illustrates how important it was for us to carry out a detailed reconnaissance in order for the IKP to avoid all areas of possible conflict before making an official application to the Pakistan government.

Since the partition of India in 1947, the problem of Kashmir and the Northern Areas, the wars between India and Pakistan, the Indus river dispute, and the conflict leading to the creation of Bangladesh, have all contributed to making access to the Karakoram somewhat difficult. Pakistan has also experienced a degree of political instability, with frequent changes in the structure of the government. In addition, the bureaucracy is such that decisions are hard to obtain. In the armed forces and in the upper ranks of the civil service there are excellent administrators, but the lower echelons are less effective, and there is a shortage of trained personnel with the ability and self-confidence to take action on their own initiative; consequently, all matters have to be referred to the top with the result that it is almost impossible to get a quick decision. In my view, if Britain really wishes to help the developing nations, we should train within our schools, polytechnics and universities the civil administrators so desperately needed by such countries, rather than introducing legislation that decreases the flow of overseas students.

In the criticial 1960s, Pakistan turned towards its neighbour China for economic assistance, and so it came about that these two nations planned and constructed a road that would eventually join Karachi on the Indian Ocean to Beijing (Peking) the capital of China, via Islamabad, the capital of Pakistan. This was the now-famous Karakoram Highway. While the road was being built, all expeditions into the north-western areas of the Karakoram were discouraged. Even in relatively good times, permission to enter the Karakoram is not a straightforward affair. In 1957, for example, I was initially refused permission to join Eric Shipton's trip to the Siachen. Many letters and petitions eventually overturned this decision, and in the end I was allowed to join the expedition, which I myself had initiated while an undergraduate at Imperial College in 1955. Ascents of several peaks, two over 6,100m (20,000ft), and journeys over the longest glacier outside the Polar Regions whetted my appetite for exploration, and so on my return I started on plans for a small four-man international expedition to the unknown slopes of K12, 7,740m (25,400ft), beyond the Saltoro Valley. In May 1960, only a few weeks before we were due to fly to Pakistan and with all our stores ready for shipping, permission to enter the Karakoram was refused. Eventually, intense negotiations by the American Ambassador and the Australian High Commissioner saved the day, but the hassle, frustration and protracted knife-edge debates, all of which required a quite disproportionate input of time and energy, made their mark. Next time, if there was to be a next time, I was determined things would be different. How innocent I was.

The reconnaissance for the Karakoram Project was started in 1978 by John Hemming, the Director of the RGS, during a 14-day trip to Pakistan on his way back from the rain forests of Mulu in Sarawak (north-east Borneo). After visits to Lahore, Gilgit, Rawalpindi, Karachi, Peshawar and Islamabad to consult universities and civil and military authorities and to seek personal links with those who could best help us, John produced a detailed twenty-six page report. Re-reading this report today, one realises just how much was achieved in so little time. Of particular importance was the list he drew up of problems involved in despatching a major expedition to Pakistan. These included the political uncertainties in Afghanistan and Iran, both of which border Pakistan and both of which were soon to explode into war.

The major advantage of this early reconnaissance was the link made with Peshawar University and the group of scientists led by Professor Tahirkheli, Head of the Centre of Excellence in Geology. Finally, just before John left for home, he talked over the telephone with a possible link-man, David Latter of the British Council, who appeared willing to give a little time to assist the project through its possibly lengthy negotiations.

John's report to Council was accepted, and it was agreed that I should follow up with a visit to Pakistan as soon as possible, in order to negotiate the necessary permission for a series of scientific studies in the Karakoram. I was also to seek the support of all interested parties once these had been identified and once I was convinced they would be able to make a useful contribution. Finally, a host of detailed questions had to be answered concerning schedules, logistics, finance, scientific equipment, food, camping and climbing equipment, radio frequencies, transportation and fuel supplies.

Those unaware of the difficulties of negotiations in countries like Pakistan might conclude that this second reconnaissance was a total disaster, despite the voluminous report I presented to Council on my return in June, 1979, for I returned without the necessary permission ; I had no commitment from any of the government agencies, and nor did I come back with any clear picture of the contribution that could be relied upon from the various civil, military and educational establishments.

Despite all these important shortcomings, however, the trip was exceedingly successful. Firstly, I had once again teamed up with David Latter, with whom I had worked for three years in Nigeria when both of us were involved in establishing a new university at Zaria in the north of that country. We both could be described as keen outdoor types, and soon David was to be as committed to the success of the expedition as anyone could be ; indeed, without his enthusiasm, efforts and personal contacts in all branches of government, we would not have been successful. Secondly, I was now aware of the innumerable and exceedingly difficult problems we had to solve. Finally, I had two brainwaves that I knew would help establish our credentials in the eyes of those in power and also bring maximum benefits to the scientific community in Pakistan.

The first idea was to include Chinese scientists on the expedition and thus make it a truly international one. China and Pakistan were now on very good terms with one another and those Chinese researchers I had met in Britain were men and women of great talent, exceedingly highly motivated and above all very pleasant to work with. I knew they could easily be assimilated into the spirit of collaboration that would be so essential to the success of the project. I mentioned the idea to David and he agreed that a Chinese contingent would not only be a desirable addition to the team but would also be essential in helping us break through the bureaucratic barriers that now confronted us. With Chinese experts alongside us, we would be seen to be a most serious project, especially if we could achieve sponsorship from the major scientific and academic societies of all three countries.

The second brainwave stemmed from several lectures I had given to Pakistani scientists on a broad range of subjects. Once again, I was made aware that although the scientists I addressed were very able, there were too few in any one discipline to be really effective and they lacked the kind of infrastructure that would enable them to make use of all the advanced work now being conducted in the western world. This situation existed in all mathematical and technological subjects, as well as the earth sciences with which we were to become very heavily involved. In spite of their unquenchable thirst for knowledge, Pakistani scientists are sometimes reduced to despair because information that results from research programmes in Pakistan, which should be made readily available to them, is instead transported away into western society by the visiting armies of teachers and researchers. This is particularly true in the subject areas of geography, geology, botany, archaeology and zoology. Foreign scientists who do research work in Pakistan invariably take their findings back to their own countries to be published in western journals. Regretfully too little information is fed back to those who need it most and then often too late.

I decided that we should take advantage of the fact that some forty-six British scientists were to descend on Pakistan by holding an international conference, where we could both exhibit our instruments and talents to the scientific community and also present technical papers. In this way we would be able to demonstrate what type of research could be done in the Karakoram and what we could achieve, as well as detailing the present state of knowledge in the several disciplines represented by the expedition. David Latter thought that this was an admirable plan and, as the representative of the British Council in Pakistan responsible for British educational aid and for scientific and technological collaboration and exchange, he felt that such a conference was one to which the Council should give its fullest support.

Returning to London with these two plans in mind not only tempered my lack of success in other directions, but also fired me to a high peak of enthusiasm. I knew that we could not fail. Furthermore, in David Latter and his wife Alison and his staff at the British Council, we had staunch allies in the struggle ahead. The Council of the RGS accepted my fifty-five-page report and agreed to press on.

Another benefit of my first reconnaissance was that I was now aware that we were in for a protracted struggle before the expedition would be given an official blessing. For purely mountaineering expeditions, the procedure of applying to the Ministry of Tourism and Culture for permits to enter the locality is relatively straightforward, but still difficult enough. For our own project, it appeared that we would have to deal with more than ten Ministries, as well as the Military. With regard to the latter, fortune smiled upon us when I met Major General Safdar Butt (currently Head of FWO) whom I had met in 1957. Soon it was agreed that I should give an evening lecture to the Alpine Club, one of whose officers was Daud Beg, whom I had been friendly with when we were both students at Imperial College. Back in those days I had initiated the College Exploration Society, with assistance from the academic staff, including Alfred Stephenson of the RGS Survey Committee. Daud had been a Liaison Officer on a Karakoram expedition and still retained fond memories of the college, and I knew he would be an invaluable friend – especially since he could vouch for my credentials, having coaxed me to give several lectures to his team of design engineers in both Lahore and Islamabad.

Excellent contacts were made at the University of Peshawar and the University of the Punjab, but a more diffident stance was taken by the Water and Power Development Authority (WAPDA) in Lahore, who clearly wished to know how far we had progressed with our application for permission from the government. Their caution was to be echoed by other organisations, and it underlined the need for careful step-by-step progress. WAPDA had important gauge stations on many rivers of the Northern Areas and gave me several reports to take back and analyse in the UK. Without doubt they would be useful allies, and we in our turn could produce technical data of great benefit to them. I felt we could probably count on a small, self-supporting unit from WAPDA joining our group. They could instruct us on recent studies on irrigation developments, and we could bring along modern water-sampling systems for their inspection and use. Together, we could provide useful information on the characteristics of the Indus and its tributary waters, and on the load of sediment that was continually being deposited into the Tarbela Reservoir.

Several visits to the Survey of Pakistan (SOP) once again highlighted the urgent need for us to get official permission for the expedition. It was obvious that, however enthusiastic public bodies might be, no one could further our plans until the slip of paper arrived from the Pakistan government giving us the go-ahead. The SOP gave me some details of available maps and aerial photographs, but I was not able to obtain copies of any documents until permission arrived. However, their willingness to talk over our plans did shed light upon several aspects of the proposed survey programme that required expansion. They were particularly interested in accurate positioning of new mountain stations with the aid of satellite doppler systems, and in expanding their own triangulation network to the east of Hunza. Their major objective was to use our presence to extract as much data as possible for use by their own field teams over the next decade – an objective with which I was in full agreement.

I faced a more difficult situation when I went to see the Geological Survey of Pakistan (GSP) in Quetta. In the immediate past they had given much support to so-called international scientific expeditions, providing lorries, jeeps, fuel, tents, manpower, food, equipment and scientists, as well as much of their valuable time.

Unfortunately, they had been repaid with little feedback and even less credit, especially where the authorship of technical papers in the world's leading journals was concerned. Naturally they were therefore very concerned to know what we proposed to do. The fact that the RGS was to mount perhaps its most intensive technological and scientific project ever could not help them overcome their own political problems, concerned with their relationships with national institutes and government bodies, or create enthusiasm among their staff for our project, even though it appeared that the Ministry of Petroleum and Natural Resources would become our sponsors in Pakistan.

Much later I learned why the GSP had to be wary of international co-operation. Before my arrival, a powerful team of seismologists from the USA had been requested to pack up their instruments and leave for home within twenty-four hours. No explanation had been given. The frustration of both Pakistani and American collaborators could be understood; after many months of planning, when all their instruments had been finally ready to take recordings, some in difficult terrain, they had been forced to retreat without any results.

In the Lourdes hotel that night I talked to an ex-British Director of the Geological Survey, who was having a similarly frustrating time trying to give money for the preparation of geological maps – in spite of having life-long friends at GSP to help him untangle the bureaucratic red tape. I spent the night squashing beetles and reading an MA thesis entitled 'The influence of Nietzsche on Eugene O'Neill', composed by the daughter of one of the senior GSP staff. It was a subject about which I knew as little as the subtleties of government departmental politics.

Back in Rawalpindi, Major General Butt had called a meeting with representatives of two Ministries, those of Tourism and Culture, and of Petroleum and Natural Resources, at which it was finally agreed that our plans should be submitted through Mr A. M. Khan of the latter Ministry, which at that time seemed to be the most appropriate one to handle the matter. However, government elections were due to be held in November, and we were advised to wait until these were completed in order to avoid losing out in the realignment of power that would follow. I could foresee that this would create a difficulty back in London, where some people were now, understandably, beginning to feel a little more dubious about the whole plan. Little did I realise that far worse was to follow.

During my next visit to SOP I met Qureshi, who had been attached to us as a surveyor in 1957. He was one of the few trained staff left, most having long since departed for high-salary posts in the oil-rich states of the Middle East. It appeared we would have little opportunity of acquiring trained assistants from the SOP during the course of the expedition.

And so my first reconnaissance came to an end. Despite the obvious problems, I could see the way forward now, and I had made some valuable friends who could appreciate what we were attempting to do. The question now in my mind was whether I would gain the support of those who mattered back home. Fortunately I did. John Auden, Chairman of our Committee, had spent some time at Quetta with the then Survey of India, and Lord Hunt had climbed in the Karakoram while a serving officer in the Indian Army. John Hemming, too, had few illusions about the difficulties we faced. It was agreed that we should postpone our application, the general format of which was at last settled, until after the November elections in Pakistan.

Planning went ahead, somewhat nervously it must be admitted. In the meantime, I had a major international conference to organise at Cambridge in August, followed by another immediately afterwards at Sheffield, and these kept my mind off numerous nagging questions. Project brochures were printed, and applications for finance drafted and presented to the various research councils in Britain.

November came, but the election in Pakistan was postponed indefinitely. The military government retained power, and new negotiations were being held with town, village and country chiefs to seek an alternative form of national democracy and government. I realised the time had come to return to Pakistan to seek the government's agreement, without which the RGS could not give its own final stamp of approval.

Meanwhile we had moved forward in other directions. It had been agreed to include Chinese members on the expedition, and letters had been despatched to the Academy of Sciences in Beijing giving full details of our proposals. It had also been agreed to hold a conference in Islamabad as soon as the team assembled in Pakistan.

I returned to Pakistan knowing that David Latter had discussed our plans with many Pakistani officials and had presented our application for approval to the Pakistan government. This went via the Ministry of Science and Technology, who it was now thought were the most suitable agency to understand and vet our proposals. I was to leave for Beijing in three days, and I had hoped to be able to exhibit to the Chinese the approval of the Pakistan government. However, various meetings at the University Grants Commission (UGC), the Quaid-i-Azam University, the British Council, and the British Embassy made it clear that I was being too optimistic. A major reason for the delay was that instead of only one Ministry being involved, all the following were now being consulted : Foreign Affairs, Internal Affairs, Intelligence, Science and Technology, UGC, Petroleum and Natural Resources, Education, Culture and Tourism, Northern Areas, Information, Defence, as well as the Military. It seemed there was no hope of permission being formally granted before I went to China, but I was told the situation looked good and that I could expect a favourable response upon my return.

I travelled to Beijing in confidence and with a stomach pumped full of Streptomagma to overcome my rebellious digestive system. On 24 February I awoke from a stupor induced by lack of sleep, just in time to peer out of the plane and see the highest peaks in the world caught by the morning sunshine. K2 looked magnificent, still not dwarfed even though the aircraft was flying at 11,000m (36,000ft). Unfortunately the Karakoram and the mountains to the north soon clouded over and I could see nothing of the area through which Shipton, Younghusband, Hayward, Gardiner, Elias and so many others had battled to extract its secrets (see Appendix I).

In Beijing's modern airport I was met by the Director of the Institute of

Mechanics, who was to be my host for the hectic eight days I was to spend in China. On Monday, discussions were held at the Academia Sinica, who agreed to select the individual scientists who would work with us on the IKP. That evening I tasted my first authentic Peking duck while being fêted by Professor Chin Li Sheng of the Academy, who led an almost unending round of toasts on subjects ranging from international science to the perpetual beauty of mountains. During the next four days I gave some twelve technical lectures and held countless discussions on topics of interest to mechanical, aeronautical and nuclear engineers. I also visited laboratories and witnessed the rebirth of scientific endeavours after the ravages of the cultural revolution. I felt a deep comradeship with those researchers tackling the same problems as our own people in the UK, and the pleasure of the visit was heightened by their enthusiasm for the IKP and their desire for it to succeed.

On Saturday, 1 March, I held four meetings. The first was with five geographers associated with the Academy, including the General Secretary of the Chinese Geographical Society, Qu Ning Su, the leader of several Tibetan expeditions, Sung Hong Lei, and Li Wen Hua and Guo Sao Lei, who were experts on natural resources. To conclude this meeting, I gave a slide show of my 1977 expedition to Vatnajökull in Iceland. The next meeting was with the first successful Everest climbers from China, geologist Wang Fu Chou and forestry worker Chu Ying Hua, the latter having lost his toes in the 1960 success after a thirty-six-hour climb, of which five hours were spent overcoming the difficult second step. This pleasant meeting ended with a request for me to show the slides of my 1975 traverse of the Staunings Alps of East Greenland. Immediately afterwards I met Li Shu-Bao, Wang Wenying, the Deputy Director of the Lanzhou Institute of Glaciology and Cryopedology, and Yang Bing Ping, Assistant Professor from the Institute of Geology. Dr Wang informed me that it was he who had found the remains of Dr Warburton, who had been lost in an avalanche on the Batura Glacier in the Karakoram. All had obviously been consulted and were well-primed about our plans, and so were able to make progress on how we planned to co-operate during the course of the project. Arrangements concerning logistics, finances, equipment and schedules were concluded within an hour, but all depended, of course, on the eventual approval of the relevant government agency and the Academy of Sciences. The latter, however, had already informed me that they anticipated no major difficulties.

We concluded our meeting by a slide show of my Karakoram visits in 1957 and 1960. The speedy negotiations in China, which all took place within a week, were in stark contrast to the protracted discussions going on in Islamabad, which were now reaching a crucial stage.

After supper, Professor Lei Tianjue arrived with Mrs Zhou. This kindly elderly man and his cheerful companion remembered their visits to my laboratories in Sheffield, and we talked in a most relaxed atmosphere about those topics that interest most engineers' minds, but to which we can devote all too little time in our busy world. This final meeting of the day epitomised what most research work is about, whether it is conducted in the world's highest mountains or the controlled conditions of a university laboratory, namely, to find routes to solutions of long-standing problems. If this can be achieved by forming new friendships across national boundaries, then this is a more than welcome bonus.

That night, in an enthusiastic burst of energy, I brought my diary up to date. On Sunday I visited the Great Wall and the Ming Tombs. On Monday I saw the Forbidden City, the Zoo and the Great Square, and then flew back to Pakistan immediately after lunch.

On my return to Islamabad I was informed that permission had been granted for the expedition, but only as far as Karimabad in Hunza. This was a major setback for several reasons. Firstly, after my discussions in Beijing, it was obvious that the Chinese wished us to operate inside China so that we could complete a network of seismic stations surrounding the Karakoram. Accordingly we needed to have access to the Chinese–Pakistan border at the Khunjerab Pass. Furthermore, in our applications for research grants we had stated that we wished to carry out several wide-ranging experiments, and this we would not be able to do if we were confined half-way up the Hunza Valley. The hallmark of approval for an expedition with an international membership would be the freedom of its members to move across international boundaries, albeit within prescribed limits. Although all travellers other than Pakistani and Chinese nationals were stopped at the Karimabad checkpoint, for our expedition to be stopped half-way up the Hunza Valley made no sense at all in scientific terms. It was important to inform our Pakistan subcommittee (Pakistani consultants and some participants) of my deep concern and to elicit their support for an attempt to gain access to the border at the Khunjerab Pass.

At the meeting held at the UGC I reported on my successes in Beijing, but it was obvious that the Geological Survey of Pakistan were still on the defensive. Some progress was made at the Survey of Pakistan, and I was able to collect a few more details of the available aerial photographs, including flight-path data, maps and costs. A trip to Lahore led to lectures at the University of Engineering Technology concerned with the safety of engineering structures, particularly high-temperature plant for the generation of electricity, including nuclear power stations. This was followed by a visit to the Water and Power Development Authority. Both these contacts were later to be of the utmost importance to the Housing and Natural Hazards group and the geomorphology team. I now had the approval of both organisations, to supplement the accreditation of the Pakistan government, and so letters could be speedily despatched to our other partners and consultants in Pakistan, including the Directors of the GSP and SOP.

On returning to Islamabad I went immediately to the Latters' house, which was fast becoming a second home. Here, on 7 March, only six days before my departure for the UK and thirteen before the final and most crucial meeting of our expedition committee at the RGS, I was informed that permission to proceed had been withdrawn. Unbelievably, despite all my lengthy discussions, at some considerable personal expense and effort, all had been in vain. The despair I had felt in 1957 and 1960 was nothing in comparison to what I felt now. I racked my brains to determine what I had done wrong, but nothing came to mind. I wondered who it was who did not fully comprehend what we were about, and what considerable benefits in terms of science, technology and education would be lost to the academic community of Pakistan if the expedition did not take place. Was there any connection, however unlikely it seemed, with recent events such as the seizure of the hostages in Iran, the attack on the American Embassy in Islamabad, or the invasion of Afghanistan by the Russians? Whatever the reason or reasons, I had to get this devastating decision reversed before the fateful meeting on 20 March in London that would otherwise undoubtedly cancel the project.

A key figure in our negotiations with the Pakistan government was Dr Afzal, Head of the UGC. Before going on to discuss with him the development of engineering technology in Pakistan universities, on which I had been asked to prepare a report assessing the situation in those areas requiring immediate development, I told him of our present plight and asked him for his support. I also went to see the major figure responsible for preparing our case to the government, Dr Manzoor Ahmed Sheikh, the Joint Secretary of the Ministry of Science and Technology, and told him of our delicate situation, stressing the need for immediate positive action and assuring him that if he could help us we would be forever grateful.

The British Ambassador, Oliver Forster, also proved equal to the challenge, and this friendly, perceptive and sincere man took our case to the Ministry of Foreign Affairs. The British Embassy, now fully briefed on our proposals, could see both the short- and the long-term benefits that our project could bring to Pakistan, and it was clear that the entire staff at the Embassy would give us every possible assistance. At the slide lecture at the Alpine Club in Rawalpindi that evening, more than one VIP came along and privately stated that there had been a misunderstanding somewhere and that I need not worry. To these and others I expressed my thanks for their support, but told them that if I did not have an agreement in my hand by the time I left Pakistan, the RGS would have no option but to cancel the expedition. This would be an unfortunate and unpleasant step, but one with which I would have to concur.

My visit to Peshawar was now shortened to one day, but it was enough to show me that Tahirkheli was obviously keen to continue planning and he had now increased his own contingent from four to ten members. On my return, I learned that discussions had been held between the Chinese and British Embassies in a further endeavour to unlock the mystery, and I was told individually and privately by many Pakistani colleagues in government circles that the matter was being attended to urgently and that I should consult the Ministry of Northern Areas. In a mood of optimism I continued to do our shopping by making banking arrangements and contacting local suppliers of fruit and meat. I also played cricket in a side street until besieged by scores of young children who wanted my autograph, thinking I was the great Australian cricketer. I shamefacedly acceded to their request, since they had applauded my excellent stance, positional insight and perfect style, and had chosen to disregard my inability to make contact with my bat or take wickets.

I attended several hectic meetings at several Ministries, and had the opportunity of once again insisting that we only wished to conduct a series of purely scientific studies, that we would bring no disrespect to Pakistan or its neighbouring countries, and that many useful benefits would accrue to Pakistan from our work. These meetings clearly illustrated the need for direct contact with officials, since without it it was impossible to gauge the sincerity and difficulties of collaborating parties. If Britain were ever faced with such an influx of foreign scientists, no doubt the authorities would be equally inquisitive, and I felt there was no alternative but to continue to press our case.

Day after day we were promised that soon, very soon, we would receive better news, but as I boarded the aircraft home I still had no letter, no telephone call, no word. The telex that I prayed would await me at London airport did not materialise either, and it was with a heavy heart that I started my report to the RGS committee at 3.00 pm on 20 March. Only a few moments before I finished, with the fatal vote just about to be taken, a secretary entered with a telegram which was pushed back and forth between John Auden and myself, neither of us daring to open it. At last I opened the paper and read the message, keeping one eye on those assembled. Permission had been reinstated. The smiles on all faces clearly indicated not only a unanimous vote in favour of carrying on, but a committee that was now whole-heartedly behind the IKP.

Much later, in the UK, I had the good fortune to discuss the crisis with an expatriate officer who knew the subcontinent well and who had much sympathy with Pakistan and its difficult political and geographical position in world affairs. His account of the background to the government's deliberations was as follows:

'Right from the start the project was a very risky one. The various people in all the different government ministries and departments that you saw all wanted in principle to say that they welcomed any form of scientific collaboration between British institutions and the government and universities of Pakistan. At the same time, not one of them wanted to take any firm, concrete steps that would actually make the project happen. This is because they all knew that if, later on, the project were to give rise to a situation causing problems or difficulties for Pakistan, there would be trouble for all those who had done anything more than make polite murmurs.

'There was good reason for apprehension. It would not have been forgotten in government circles that there had been an occasion when the Americans were believed to have used permission granted to scientists to enable intelligence agents to plant some kind of listening or navigational device in the mountains bordering on China, to the detriment of Pakistan's security and her relations with other states. The truth or otherwise of this allegation is irrelevant here; what matters is that it was believed, and that officials dealing with these events did not want to risk a repetition.

'Furthermore, although it was possible to persuade most of the people with whom you dealt of your bona fides, another characteristic of the Pakistan bureaucracy is that it works very much by precedent. Anybody considering giving permission to the Royal Geographical Society would be well aware that the next step might be a demand from other countries for similar facilities to be accorded to another large group of "scientists", all intent on getting as close to the Chinese border as possible. Perhaps, also, the government feared that they might face increased pressure from the Soviet Union in the wake of the invasion of Afghanistan, on the grounds that the government had allowed foreign agents to operate close to the Russian

27

border. This *canard* was indeed raised later on in the Indian press, in articles clearly inspired by the USSR.

'Thus everyone wished to see the decision taken elsewhere. But after much careful diplomacy behind the scenes, it did appear that the best chance of success lay in getting the Ministry of Science to take on the co-ordinating role from the Pakistani side. This would emphasise the purely scientific aspect of the project and thus allay the fears of other departments. Even here, another aspect of Pakistan bureaucracy came into play : that permission for anything to happen is not so much a question of one competent body deciding that it ought to be done, as of reaching a position where every single department that can conceivably be involved has been consulted, and says that there is no reason why it should *not* happen. Pakistan in fact works on the "No objection certificate".

'The Ministry of Science did its best. The appropriate people were consulted, and approval was issued. But after that, doubts arose. Perhaps some of the departments to which reference had been made had raised objections which could not come to public knowledge. There may have been concern about the wisdom of having scientists with sophisticated equipment anywhere in the Northern Areas at all. If there were these objections, they probably had their origin in the past, allied to the fear of setting a precedent for the future, rather than in the project as such. Be that as it may, permission was withdrawn, though officials in the Ministry of Science did their utmost. Fortunately, in the end those friendly to the project were able to reassure the doubters and in the nick of time the original permission was reinstated. Perhaps the one lesson that does emerge above all others is the supreme importance of personal relationships. In the last analysis, it is not our detailed understanding of procedures that will help us nor our skill in presenting arguments and assessments of the gains and benefits ; what matters is how we are regarded as people – as friends whose goodwill is manifest and whose word may be trusted.'

And so it appeared that on the one hand the Pakistan government had been both courageous and far-sighted in permitting the Project to go ahead, whilst on the other hand it was beholden on us not to put a foot wrong and under no circumstances to embarrass them. After all, as well as the episode mentioned earlier, the Pakistanis would remember the so-called 'CIA plot' in India in 1962, when scientists on an American expedition to Nanda Devi were reported to have planted a nuclear-powered device on the mountain to monitor the Chinese nuclear programme. It was said that although much preparatory work had been done on Mount McKinley in Alaska, the device did not work when it was planted on Nanda Devi and it was subsequently lost in an avalanche. Since the snows of Nanda Devi feed the headwaters of the holy river, the Ganges, the seriousness of the incident can be understood.

Perhaps it was this background that caused *Pravda* to list us as a CIA-backed expedition, which was of course totally unjustified; it may also have influenced the attitude of the Pakistan government in their initial anxieties about our project, until it became clear to all that we were exactly what we said we were.

Let us hope that in the not-too-distant future scientists of all countries can combine their talents to work in this area, where nature alone has erected enough barriers to exploration and scientific endeavour without the need for further complications.

# The Last Lap

It was with no little apprehension that I returned to Pakistan on 14 June 1980. I fervently hoped that nothing could go wrong now, as the entire expedition was soon to follow. We had done our homework and I felt there should be no misunderstanding about our desire to achieve good research results in a spirit of international co-operation. We had many objectives, perhaps too many, but with only ten to twelve weeks in which to complete a series of investigations we were keen to get started. Personally, I wanted to be involved in the glaciological and survey programmes, and somewhat selfishly I hoped to be able to devote a high proportion of my time to these studies. Quite simply, I now wanted to see, live and work for a short time in the Karakoram – mountains that for so long had been only a pleasant memory. The following brief extracts from my personal diary indicate how some of the details of this, the last and probably most important stage in the preparation of the expedition, developed and crystallised to set the environment in which we would have to work.

'I am exceedingly disappointed that after so much preparatory work little seems to have moved at this end. Some Pakistani members have withdrawn, to be replaced by new members. I am putting Professor Tahirkheli in charge of co-ordinating the contributions of the new Pakistani scientists. Permission to go further north than Karimabad has still not been given, although the Chinese Embassy require full details of when we intend to operate across and inside Chinese territory. At the subcommittee meeting, the representative of the Ministry of Science and Technology was absent, and I was later told that he had not been given permission to attend. Fortunately, Dr Afzal [the Head of the UGC] took the chair and was seen to be a man of the highest integrity and ability . . .'

'Nigel informed me that no contact had yet been made with our two Liaison Officers and that the railway truck containing all our stores was lost somewhere between Karachi and Rawalpindi. To everyone's intense relief it was found three days later . . .'

'David and Alison Latter of the British Council have done wonders. All members of the Project, on arrival in Islamabad, are to be housed with British expatriates. My schedule of lectures, visits and receptions has been prepared. Meanwhile, David has accompanied Nigel to Karachi, Quetta and Lahore to assist the clearance of our baggage and liaise with Pakistani institutions . . .'

'Much is going on around me and I am not yet able to draw in all the reins, but this does not worry me, since satisfactory and precise decisions are being taken by Project enthusiasts and this leaves me to deal with the problems of getting permission to go beyond Karimabad, the new members and their questions, and the conference . . .'

'At the UGC meeting on Monday, Afzal stated he was to see the President of Pakistan, General Zia-ul-Haq, and that he required detailed information regarding the forthcoming conference. It was hoped that the President would agree to inaugurate the conference and so give the stamp of approval to the Project. Afzal agreed to bring up my request for siting our

29

base camp at Pasu, some 50*km* (31 miles) beyond Karimabad, and asked me to prepare an aide-mémoire on the setting up of a "Karakoram Research Cell". Finally he queried why WAPDA had now withdrawn their support and agreed to follow this up privately in an attempt to induce them to have some input into the Project.

'At the Ministry of Science and Technology, I met a nervous Dr Kazi, our Contact Officer at the Ministry. He obviously did not want to make a mistake and in this respect he had my sympathy, since if anything did go wrong he was the man who would be first in the firing line. However, Dr Manzoor Ahmed Sheikh, the Permanent Secretary at the Ministry, gave approval for the release of air photographs of the Hunza valley (although only as far as Karimabad), which indicated that another major hurdle had been overcome (or so I thought). In the late morning I had the pleasure of meeting Maj. Gen. S. Shahid Hamid, Minister of Information. We talked affectionately of Trinity College, Cambridge and as I left he presented me with his book on Hunza, written whilst he was a close friend of the late Mir, Jamal Khan . . .'

The President of Pakistan did graciously agree to inaugurate our conference and on 21 June Lord Shackleton flew in from the UK to represent Britain in the ceremony. His trip to Pakistan covered a very hectic thirty-six hours during which he proved to be as dynamic and lovable a character as his father, who was possibly the greatest of all British Antarctic explorers. Being a Knight of the Garter and having been on several expeditions himself, his visit was exceedingly important since it not only emphasised the significance of the Project to the Royal Geographical Society, of which he was a past President, but also indicated a degree of approval by the upper echelons of the UK establishment.

Immediately after he had stepped off the aircraft and been greeted by Oliver Forster, the UK Ambassador, amidst a battery of TV and film cameras, we went into a press conference. Questions ranged over the whole spectrum of possible commercial and scientific aspects of the expedition, though we also faced some impertinent questions from a rather aggressive young American AP reporter, who constantly enquired about our movements close to the Afghanistan and Chinese border and did his best to construct a non-existent story concerned with the political and military implications of our expedition. We took the opportunity presented by the press conference to point out that the Karakoram is not alone as a zone for hazardous living, quoting as examples the Aberfan disaster in Wales and the risk of flooding in the centre of London if the London barrage system was not completed quickly. Once the business was over, a most enjoyable few minutes were spent welcoming many new and old friends, some of whom had accompanied me on previous expeditions to Iceland and Greenland. At long last I felt we were about to get down to business.

After a short briefing session at the Ambassador's house concerning our hurriedly arranged evening's visit to the President of Pakistan, we left for the Latters' house, where Lord Shackleton met all members of the expedition.

The visit to General Zia in Rawalpindi later on that day proved to be another turning-point in the attainment of our objectives. Introductions were effected by Oliver Forster, and we spent a few minutes discussing local affairs and the British political scene before we turned to expedition matters. In an increasingly relaxed atmosphere I gave a very brief summary of the purposes of the expedition and emphasised the desire of both the Chinese and British scientists to collaborate with Pakistani scientists in both University and government establishments. The President was obviously well-briefed and knew of our programmes in some detail. We summarised the difficulties we were having in obtaining permission to work in the Northern Areas, while he outlined the delicacy of the problem in that sector. When he remarked that he would give favourable consideration to our requests, I jumped the gun somewhat and thanked him for his sympathies in these matters. The Ambassador gently pointed out that the President had not yet given permission, but would review the situation. But after a few more issues had been debated and the President was totally aware of the involvement of and agreements with the Chinese, he agreed to unfurl the three nations flag, designed by Nigel Winser, during the conference inauguration ceremony. This further encouraged me to express the expedition's thanks, which in turn may have persuaded him to take the matter one step further, because he then stated that he would expedite matters, consult with other bodies concerning our request, and if at all possible assist in granting the permission we so eagerly awaited. We drove back to the Embassy feeling much relieved.

That evening we had an excellent dinner party at the Embassy, attended by all the important Chinese and Pakistani administrators and politicians. It had been agreed the previous evening that if the weather was fine on the morrow a party including Lord Shackleton, the Ambassador and his daughter, David Latter and myself would fly to Gilgit. We eventually took the early flight after many doubts and telephone calls to Gilgit to check on the poor weather conditions. Paul Nunn, Nigel, Shane and David Giles were also going through to Gilgit on the same flight, to act as an advance party. Shackleton was impressed, as were all members new to this experience, especially when the fragile plane flew past great crags and over very narrow cols, skimming the trees at altitudes of approximately 4,500m (14,750ft). Because of bad weather, we followed the Indus from above Tarbela Reservoir and did not fly the Kaghan Valley and Babusar Pass route. Although only parts of Nanga Parbat were visible, the snow, ice and rock bastions of the upper 3,000m (10,000ft) of the north face could be seen drifting through gaps in the clouds above us, while another 3,000m below was the Indus Gorge, weaving like a snake in the heat of the desert.

An amusing conversation took place as we flew over the last ridges into Gilgit. With the prime aim of causing the maximum apprehension to the Ambassador's daughter, the pilot asked his co-pilot, 'Can anyone see the airstrip?' We looked in all directions. 'Has it gone?' We detected a hint of seriousness. 'It must have been moved, but where to?' No reply. 'Has it been covered over by a landslide? Perhaps it's been camouflaged due to the Afghan troubles.' The tiny airstrip was only seen at the last possible moment, when the Fokker Friendship banked steeply, flew parallel to the silhouette of the sharply descending ridge and then, turning out of a tight circle, dropped onto the tarmac, missing several prominent crags on the way down. It was quite an experience to fly without computer aids and with natural eyesight as the only navigational equipment. All of this amused Shackleton immensely, but I wondered what effect it must have had on the young girl seeing harsh mountain country for the first time.

Unfortunately the weather was so bad that we could not see the peaks of K2, Rakaposhi or Haramosh. On the runway it was obvious that we should return from Gilgit as soon as possible, since angry clouds were drifting in from the west, and the Ambassador, Latter and myself all became a little nervous that we would not be back in Rawalpindi in time for the inauguration later that day. At this point Shackleton was still keen to visit the hill station at Murree, but as he returned to our group I told him that there had been a conspiracy and that under no conditions would any of us drive him there for fear that he would not be back in time to deliver his conference address, alongside the President of Pakistan. We were all very grateful when he saw our determination and reluctantly agreed to our suggestion.

It was the first time I had been into Gilgit, and so took great delight in examining its extraordinary position. Although there are hidden exits to the north, east, south and west, it appeared to be situated in a deep bowl with no possible way out – as, indeed, is almost the case. Once in the little green oasis in that arid desert, few would want to stray from the limited protection it offers, and only the inquisitive or unwary would dare to probe the deep gorges and high passes to seek other equally well-hidden oases many, many kilometres away. At the airport it was good to see Bob Stoodley, our Transport Manager, who had already explored the road and tracks a long way up the Hunza Gorge in our British Leyland Land Rovers.

Back in Islamabad, the inauguration ceremony was a tremendous success. The two introductory speeches were followed by one from the Minister of Education, Muhammad Ali Khan, all concerned with the importance of the international conference and the project as a whole, and with the necessity of mounting this kind of venture again in the future. The President spoke in Urdu, cracked a couple of jokes and indicated his desire to see us succeed in all our endeavours; Lord Shackleton then followed with an impromptu and informal speech, brief but very much to the point. Immediately after the final introduction by the Minister of Education, I gave a keynote paper entitled 'Some Recent Technological Advances in Earth Sciences', in which I dealt with the power of steam, water and ice, since the glaciers and rivers of Pakistan are a major concern of those responsible for understanding and developing the natural resources of that country. During the interval, General Zia made it clear that he wanted as much interaction as possible between his own scientists and those from China and the UK and that every effort should be made to attain this collaboration. This was seen to be one of the more acceptable faces of international technical assistance. The unveiling of the international flag by the President and myself was the closing event of a very busy day, and I hoped it would be interpreted as a symbol of our unity of purpose.

After the opening of the conference, Lord Shackleton retired for a few hours' sleep before flying back to the UK, apparently unaffected by his strenuous thirty-six hours in Pakistan. After this foretaste of the endurance and abilities of senior citizens from the House of Lords, we began to wonder what standards Lord and Lady Hunt would set when they came to visit us. The conference itself, held between 23 and 25 June, was a great success. It permitted us to exhibit all our equipment to anyone who wished to examine it. Naval officers from Karachi came to look at the satellite doppler equipment, while staff of the Survey of Pakistan and the Geological Survey of Pakistan took the opportunity to study and handle all our advanced instrumentation. The Chinese in particular were very anxious to learn of our radar impulse ice depth sounder, and many lectures were given by our experts outlining not only the novel features of this equipment, but also the full specification in terms of range and resolution. At the end of the conference I was assured that all questions relating to these scientific instruments had been answered and that no doubts about them now remained.

I opened the first technical session of the conference (details of which are given in Appendix IV, p. 199) in order to set the style of discussion. Half an hour was allocated to each technical paper, the last five minutes being intended for points of clarification. All lecturing facilities were available, including overhead projectors, screens and blackboards, although unfortunately the tightly-packed room, the high temperature, which peaked at about 35°C (95°F), and the humidity combined to beat the air-conditioner. During the first day, some thirteen papers were presented by delegates from four nations, and I was very pleased to see that discussion and criticism ranged freely, despite the language barriers and inhibitions caused by the author's reputation in the field. It was particularly interesting to see two geologists from Pakistan, Ghazanfar Abbas and Qasim Jan, fiercely arguing over a proposed model for the Northern Areas of Pakistan, and to note that the British and Chinese delegates were prepared to enter into the spirit of these discussions and complicate the issue even further.

It was a very successful opening day, but a tiring one. We started at 8.00 am with a tea-break at 10.30 am, followed by a one-hour lunch at 1.30 pm. After a coffee break at 5.00 pm, more time had to be allowed for discussions, but by 7.00 pm all members were wilting a little. The conference committee had laid on a party for each evening, the Monday reception being given by Attock Oil Company. It was with great pleasure that I met old friends from Morgah, who reminded me how, in 1957, Shipton had insisted I learn to swim. In fact he had given me precisely three days to do it and threatened to send me home if I failed. The reason for his insistence was that the ability to swim gave expedition members a slight chance of survival should any of them fall into the raging torrents of the Karakoram rivers. Many friends helped, and by the third day I managed to splutter along the length of the pool, although I admit that my feet might have touched the bottom as I came to the last few metres.

The second evening's reception was given by Habib Bank, holder of the expedition accounts. Thereafter we had much improved service. On the third evening a reception was held at Flashman's Hotel in Rawalpindi by members of the British delegation. Here I exhibited a few slides taken on my 1957 and 1960 expeditions to the Karakoram. All of these occasions were attended by many influential Pakistani civil servants, who subsequently became our friends. Most had not visited the high mountains and so became more appreciative of the beauty of their own country and our desire to travel there.

It was during the second day of the conference that we met yet another difficulty. I had asked Professor Tahirkheli to call together all our Pakistani scientists so that we could discuss future plans. I informed them that I had not yet received full CVs and information concerning their personal schedules, field programmes and financial arrangements. I explained that if VIPs such as the President and the Army Generals were to visit our base camp, as seemed likely, it was imperative that I knew where in the Karakoram everyone was, what type of work they were doing and how they could be contacted and consulted. I also reminded them that the British contingent could not meet all the expenses of their programmes and that they would have to provide their own transport, tents, food and equipment. All this had been made perfectly clear from our earliest discussions in June 1979. A prolonged discussion followed, but I was not prepared to allow several issues concerning events long since passed to influence our own commitments at this late stage. Detailed negotiations with the Pakistanis had not been possible in the previous year and we had to concentrate on planning, based on the position as it was now.

Eventually it became clear that as yet no member had contributed to the expenses of the expedition, and while this was understandable in the circumstances, it was still far from clear why not one single rupee was in the kitty for the Pakistani fieldwork. My view was that first we should agree on the total cost of their involvement and then see how, where and when the cash would be raised. I then left to attend to other issues, and on my return learned that many members wished to resign from the programme. This was just not permissible at this stage, so I suggested that we should go and talk to the most senior Minister who had so far supported the expedition. Several Pakistani members were a little nervous of this suggestion, but Dr Afzal granted an interview and a small committee made up of myself and four Pakistanis drove off at speed to the University Grants Commission.

At the UGC, Dr Afzal gave a clear picture of the background to the expedition from the Pakistan viewpoint and reassured us that while there appeared to be some mismanagement, this should not interfere with the development of the programme, and that ways and means would be found to support the Pakistani contingent. He quite correctly omitted historical questions from the discussions, as I had done, and reminded all concerned that it was to the benefit of the scientific community of Pakistan that the whole venture go forward. Eventually agreement on budgets was reached, and we returned to the Quaid-i-Azam University much encouraged.

That evening Dr Afzal telephoned to tell me that 125,000 rupees had been allocated to the University members of the expedition, to assist them with their expenses. Professor Tahirkheli was now a much relieved man, and his gigantic smile returned. This enthusiastic scientist had my full support and sympathy. He had already put in a lot of work to make the project a success, but was obviously having difficulty in co-ordinating and administering the Pakistani programme, due, no doubt, to regional differences and possibly, too, to academic rivalries between several government and university establishments representing the Pakistani contribution – as also occur in the UK. Despite all the frustrations, however, he was always willing to try to overcome problems. He agreed to continue with the organisation of the Pakistani contingent and to let me have full CVs of the Pakistani participants within a short time. Meanwhile, the conference was progressing at an encouraging pace, and I was informed that in my absence much discussion was taking place across traditional academic boundaries, which, after all, was the *raison d'être* of the whole Project.

On the third and last day, another serious problem occurred concerning the proposed work of the British scientists inside China. In London it had been decided that, rather than carry out negotiations by correspondence with all our Chinese contacts in Beijing, Islamabad and London regarding schedules of movements from Hunza to the Chinese province of Sinkiang, it would be best to finalise arrangements in Islamabad. We knew from experience that there would be minimal delay from the Chinese, and this would give us a little more time to work out precisely where we would like to place our seismograph stations. Rather than simply request those locations that gave an optimal geometric distribution, we first needed to be sure which sites it would be possible to reach in the field, using our own resources. Accordingly, I went to the Chinese Embassy, a building as imposing as the Russian Embassy on the neighbouring compound but with a much less restrictive atmosphere, where I gave Mr Liu, by now a friend of the Latters and the expedition, correspondence relating to the arrangements previously agreed in Beijing. I had wanted to change two features of these plans, but it did not seem wise to alter the names of the four Britons whom I had already mentioned as willing to work inside China (Walton, King, Jackson and myself), despite the fact that we now had the services of Bishop and Allen, who would be better employed there than either Walton or myself. As a result, instead of four people moving across the border as originally intended, only Walton, King and I remained to do this work, as Jackson was committed to another area.

A further problem was concerned with the number of trips to be made across the border. Nigel Winser had intimated to the Chinese that four visits would be adequate, whereas it was clearly necessary to cross the border every second day in order to change the graph paper on the seismic recorders. Subsequently, we agreed that thirty trips would be requested for approval by Beijing. One further issue had to be debated, namely the interpretation of '20*km* (12.5 miles) access across the border.' I assumed that this would mean a line 20*km* deep, parallel to the border, beyond which we must not go, but which would give us enough lateral movement to find suitable sites for the seismic and survey work on the far side of the Karakoram watershed, including the Chinese side of the Mintaka Pass if this was thought necessary by our scientists.

I informed the Chinese Economic Councillor that I had been told in Beijing that the Mintaka was best approached from the Chinese side, and not the Pakistan side, and that by driving down the Karakoram Highway into China one could take a track immediately to the east of the Mintaka which would permit much easier access to the area of the pass. However, one of the Chinese members of the expedition considered that this was not only too long a trek, but also too hazardous to undertake in the time available. It had previously been agreed that we would cross the border at 7.00 am in the morning and return at 7.00 pm each night; since we wished to cross the border every two days, it was obvious that these journeys would be extremely tiring to the personnel concerned. A further problem was that James Jackson now wished to work in the Skardu area, and therefore the seismographical work required more flexibility that originally planned. In the end I agreed that we would use only the Khunjerab Pass and would not, under any circumstances, cross the Chinese border, or for that matter any other international border, at any of the other passes in the area linking Pakistan, China, Afghanistan and Russia. It was obvious that the Chinese were understandably nervous of movements around this border area, but while I felt exceedingly sorry to put them in a difficult position, I was also keen to press for conditions that would allow the maximum scientific content in our programme and so help future collaboration between scientists of China, Pakistan and Britain, or indeed any other country. It was and is my strong belief that all international scientists should collaborate to the maximum extent wherever and whenever possible, as this will not only bring increased benefits to all mankind but will also promote stability and peace throughout the world.

The most encouraging feature of the long discussions with the Chinese Embassy staff and the Chinese members of the expedition was that they were obviously very keen to extract the maximum benefits from this collaborative effort; their only concern was to arrive at the correct form of wording for the telex to be sent to Beijing. The first crossing of the border was scheduled for 15 July to allow for permission from China to reach us. We estimated it would take something like eighteen days for the cable to be interpreted in Beijing, taken to the appropriate authorities for approval, and for the decision to be transmitted back to Islamabad, to members of the expedition at the Hunza Base Camp and also to the Chinese Border Guards. As someone rather pointedly stated, it would be as well not to approach the border before this date, in case the Frontier Guards mistook us for someone else and shot us.

I rushed back to the conference for the closing ceremonies, at which Professor Zhang, leader of the Chinese team, Professor Tahirkheli, myself, Dr Jafree, the local organiser of the conference, and the Vice Chancellor of Quaid-i-Azam University gave short speeches of self-congratulation. I emphasised that in my opinion the conference had been a very great success, based on my experience of other international conferences, which I attend fairly frequently. By this I meant that as many as 75 to 80 per cent of the papers were of high standard and had led to much informed discussion. It was interesting to note that the Chinese papers were characterised by a high proportion of quantitative assessments, and that their results, derived from extensive field observations in the Karakoram during the period 1974–5 and in 1978, were based on extensive computer analyses in their own universities. Three of the Pakistani papers were also of exceedingly high standard, and illustrated that despite a lack of technical facilities they too managed to keep abreast of progress and were making a worthwhile contribution to man's know-

General Zia, President of Pakistan, opening the Islamabad conference.

Rakaposhi at sunset.

*Right* A typical Karakoram track; *centre* K2, the second-highest mountain on earth; *below* A small rock-fall on the Karakoram Highway.

The summit of Shishpare above the Ghulkin Glacier.

ledge. The standard of the British papers was also high and it was pleasant to observe that our Pakistani and Chinese colleagues were prepared to criticise us over any academic weakness. Their willingness to make constructive criticisms in areas not associated with their own disciplines augured well for future collaboration in the field.

There are in fact many areas where experience in one discipline can be applied to problems in another. For example, the findings of engineers in the UK who studied the collapse of the coal tip at Aberfan, a disaster in which many children died, have applications in the Karakoram, where geomorphologists are interested in such hazards as mud flows and landslides. As a consequence of the Aberfan disaster, many experimental and theoretical studies were made of the microprocesses and macromechanics involved in that catastrophic slide. These studies produced technical data on the effects of liquefaction and other variables such as density of material, and shear strain gradients. As a result the mechanics of the fracture that caused the landslide are now well appreciated. This is the sort of engineering expertise that could well be of use to geographers. Certainly, there is today a rapid expansion in technological innovations in geographical research as this discipline evolves towards a more precise quantitative science, utilising electron microscopes and involving computer studies to derive chemical, physical and mathematical data. It was interesting to hear more of these developments during the expedition.

The conference thus came to a most satisfactory conclusion. In that period I had presented six lectures, including two on the fracture behaviour of ice and metals, the first relating to the formation of crevasses in glaciers, and the second, given at the 5th International Summer School at Nathiagali, to cracks in metals under three-dimensional stress systems as found in many engineering structures – including the aircraft that would now fly us over the Margalla hills towards our ultimate goal, Hunza.

The extension to our permit had not yet arrived, and so I had to remain in Islamabad while the main party of British and Chinese scientists flew to Gilgit immediately after the conference ended on 26 June. The fact that our permits were not yet to hand did have one slight benefit in that it allowed members to acclimatise slowly to the altitude around Aliabad in the Hunza Valley (see p. 100), which meant that they should be ready and eager to start work by the time I arrived.

My first task was to settle several outstanding issues with Dr Manzoor Ahmed Sheikh at the Ministry of Science and Technology. Before my meeting, however, I called in at the British Council offices to see the Pakistani members, who were still sorting out their logistics for travelling to Hunza. At the Ministry, I found Dr Kazi, the official dealing with our files, still exceedingly nervous when it came to discussing our permits. He had prepared our papers for advancement to the appropriate authorities, but he was still greatly concerned over concessionary details for our first Liaison Officer.

I am still unsure as to the source of our problems concerning our Liaison Officers. The previous day I had stated that it was unreasonable to make additional requests for food and equipment at this late stage, and that in my opinion it was ridiculous for us to agree to pay a daily allowance to the Liaison Officers as well as providing all their food. We had already agreed that we would provide insurance cover and free medical facilities, and subsequently we added some eating, climbing and sleeping equipment, including a sleeping bag, groundsheet and tentage. I pointed out that it was not our fault that we did not get to know our Liaison Officers earlier and their personal requirements. Neither could I understand why it was thought necessary to have two officers, the second of whom had still to be appointed.

Performing a quick calculation, I showed the administrator that for the remaining eighty days of the expedition each day's delay was equivalent to a loss of £1,000. I reminded him that, in common with all scientists arriving to work in a foreign country, we wanted to go to work as soon as possible in order to obtain maximum results from the limited period of our stay. However, in this particular instance, we had agreed to stop for several days in Islamabad, at some considerable expense, in order to give, free of charge, details of our recent research work to the international conference delegation.

I concluded by saying that, while I understood the difficulties of our position and his own predicament, it really was taxing my patience to the limit to delay our progress any further, especially since I had been assured that our permits would be ready that day. I was then ushered into the room of Dr Manzoor Ahmed Sheikh, who was having consultations with Dr Shami, the Head of the Pakistan Science Foundation. Both gentlemen received me courteously and we exchanged views on the success of the past three days.

The Permanent Secretary presented a very different attitude. He assured me that he had done his very best and explained that he was a little embarrassed by the situation because he knew what further delays meant to us, not only financially, but also in terms of lost days for scientific research. He told me that the summary papers had been modified and that, although they had still to be read by the President, we could expect a decision within a few hours. But I was asked to understand that the President was exceedingly busy with budget papers.

The Permanent Secretary had spent many years in industry and commerce and had only recently returned to government administration. It was obvious that he had command of the situation and his quiet authority, coupled with his experience of dealing with thorny problems, gave me some satisfaction. He was to become one of the three senior government officials who gave our expedition their fullest possible backing within their own sphere of responsibility. However, he explained that while he would do all he could to further our Project, there were aspects of it that were the concern of other Ministries, and he could not pronounce on their views. This certainly gave me a better insight than hitherto into the problems he faced, and I withdrew in the knowledge that we would have confirmation very soon one way or the other. Nothing, however, was said about movement up to the Khunjerab Pass.

On returning to Dr Kazi's room, I was cheered to hear that our appointed Liaison Officer, Major Rana, would not be requiring both food and a daily allowance, but was in fact prepared to live as one of us. Some details concerning our transportation into the field were quickly settled, and I was left with the impression that many of the difficulties of the past few days were now overcome.

As I returned to the Latters', I realised that the following day was Friday, the Muslim Sabbath, on which nothing could be done, and therefore Saturday would be the earliest we could receive the permits. If all went well, we should be on the road on the Sunday morning. The frustrations of the past few days were now beginning to take their toll. I paced up and down the Latters' living room nervously examining various alternative plans for the expedition in case the permit excluded areas in which we wished to do our work.

On June 28, late in the evening, Bob Stoodley, our Transport Manager, returned from Hunza. We exchanged news. Bob was extremely enthusiastic about the area and praised the work of our scientists, whom he thought at first to have been rather tired, but who as soon as their stores arrived in Karimabad had set to, opening boxes, examining and testing equipment and generally behaving as enthusiastic scientists should. I told him of the further problems in Islamabad, and of the possible reasons for the delay. Was it because the Pakistanis were unwilling to establish a precedent by giving us permits for Pasu and the border? Perhaps, therefore, they were still looking for ways and means for us to travel without the issue of permits. Perhaps, also, they were still negotiating with the Chinese – a distinct possibility if there were no frontier post and the Chinese wanted to establish a guard there at the times we wished to cross the frontier.

Ironically, however, current circumstances dictated that we could not travel beyond the point where the Ghulkin Glacier enters the main Hunza Valley, since it had recently obliterated the road. Bob also reported on two glaciers coming in from opposite sides of the Shimshal Valley, which had cut off the track and dammed the river and were causing great concern because of the possibility of a flash flood when the dam broke. Bob was also concerned that there could be a serious shortage of food in the area in the immediate future and reported having had numerous difficulties getting through check point barriers without a permit, even those that were just beyond Gilgit, up towards Hunza. It all made me rather sad and perhaps a little angry as well.

A rather amusing incident occurred to relieve the gloom. Today being Friday, I expected no communication whatsoever from the authorities, but while I was sitting in the bathroom, suffering from Delhi Belly, the telephone rang; it was Dr Kazi. Clutching trousers in one hand and the telephone in the other, I listened to his message, which was simply that he personally was much happier now, that positive movement was occurring, and that he thought it best to give me a call and so ease my frustration. I translated this message as meaning that nothing was yet in his hands and that nothing could be done tomorrow or for the next two or three days. I returned in haste to the bathroom, but could not but be amused by the incident, despite my discomfort.

On Sunday, Bob and I drove directly over to the Ministry. The situation was becoming increasingly embarrassing, but at least I now learned that the file had indeed come back from the President at 8 pm the previous evening. I then asked if it would be possible for us to travel immediately to Gilgit and for the permission to be telexed through to the District Commissioner, or as soon as the relevant papers had been prepared. This procedure was agreed. The Ministry officials were now 90 per cent certain that we would be allowed to go beyond Pasu. All expedition members were now in Aliabad and were being charged high rates for the use of the facilities there and so it was essential that we move out as soon as permission was granted. I wished to move our base camp further upstream, even if only on a temporary permit, and so promised to travel only to Gulmit (see p. 100), with the additional proviso that we would not venture further afield until the documentation had been completed. In the final analysis, however, it was agreed that the Aliabad base camp was adequate.

We then travelled down to the Frontier Works Organisation to talk to Brigadier Imtiaz. Whilst he was sympathetic, he did not think he could help us with the additional tentage we now required because of our increased membership. He also stated, quite categorically, that all arrangements should have been made months ago, with which I was forced to agree. Unfortunately, whilst these issues were being discussed during my 1979 and 1980 reconnaissances, he had been unable to take positive action since permission had not yet been granted. However, he did obtain a lorry to transport us and our kit to Gilgit. At this point, also, I was given the name of our second military Liaison Officer, Major Zaqir.

Monday 30 June was a momentous day. Dr Kazi telephoned to say he had the forms ready and could I come and pick them up. I took the first available taxi. It is worth recording that Islamabad and Rawalpindi have a taxi service consisting of hundreds, if not thousands of Morris Minor 1000s. Many of them chug along like tanks, but they are, in general, quite an efficient service. With black bodies and yellow roofs, they scuttle around like lines of beetles, collectively organised, but individually unpredictable. Over the years the interiors have become rather tatty, invariably with torn seating, but the most engaging features are numerous home-made modifications to the various instruments. The one I liked best was the dangling wire which, when touched with a screwdriver spindle acting as a transformer winding core, caused the horn to blast out a melody that frightened all animal life in the vicinity, ranging from vultures to other taxi occupants – but not, I noticed, other taxi drivers.

The service is also relatively cheap. The taxi I took to the Ministry was exceedingly hot, with no ventilation or cooling system and windows that could not be opened, so I was pleased to step out and find Dr Kazi ready to take me to the Ministry of Culture and Tourism. He had prepared a letter from his own Ministry authorising the Ministry of Culture and Tourism to act.

When we arrived we found that the Director, Nasrullah Awan, was sick and not available, but his deputy, Taleh Mohammad, a most pleasant man, was very co-operative and anxious to get things organised, being aware of the prolonged period of negotiations that must have ensued within the dozen or so Ministries before our permits could be granted. In the same room, and also seeking permits, were two Britons who had come to photograph birds; obviously they didn't know me, nor the difficulties we had faced over the preparation of CVs and security clearance for each and every member of our expedition. They stated that they wished to join the Royal Geographical Society expedition, and made several statements that could have jeopardised our precarious position. I quickly intervened to say that I did not know them and that, with deep regret, under no circumstances could I permit them to join our expedition. I also made it plain to the now-bemused official, patiently and somewhat apprehensively poised to sign our permits to enter the no-go area for foreigners, that this would apply to all other visitors coming through Islamabad who might wish to make contact with us.

During the final stage of the negotiations, the official was exceedingly pleasant and courteous, but asked Dr Kazi to make an additional note on the bottom of his own authorising letter, since he did not want to infringe some unknown rules about the transfer of responsibility for issuing permits. Both men were obviously a little nervous over this transaction, but soon the magic words were written at the bottom of the permit allowing us to set up camp at Gulmit and to travel the road between Karimabad and Gulmit. The official stamp placed on the bottom of our original permit looked insignificant enough, but this was what we had wanted for more than two years. I was elated, but upon reflection I realised that the scrawling words and official stamp in themselves were meaningless. What was important was that the officials concerned transmitted this information to the authorities in Gilgit, and that they in turn transmitted it up to the various frontier posts, security agents and liaison officers so that they would allow us to enter the areas concerned. Nevertheless, it was a momentous event after so many pitfalls, so many disillusionments, so much frustration, so much hard work and so much nail-biting.

At the British Council we duplicated the permit so that each and every member of the expedition had a copy. Additional copies were made for the Chinese Embassy, British Embassy and British Council. I then talked to Major Rana over the telephone. He had not been successful with Pakistan International Airlines (PIA) who refused to permit him to withdraw our baggage, although it included several pieces of key equipment for the seismologists which they were desperate to receive. Unfortunately, they had packed their kit into a box that could only fit into one space on one of the aircraft flying between Islamabad and Gilgit. Because of the bad weather during the past few days and the consequent rerouting of aircraft this particular plane had not flown to Gilgit. The non-arrival of the equipment had caused much inconvenience and not a little friction at base camp. Now it appeared that PIA would not permit us to take this last piece of baggage up to Gilgit by road, since it was stored in an import tax-free bond store! There are some battles that you cannot win, so I gave a display of temper and hoped the officials were suitably impressed. I had to leave this problem with David Latter.

And so at long last I was now able to return my attention to our fieldwork studies. Before leaving Islamabad, however, I had first to collect the aerial photographs from the Survey of Pakistan.

I had travelled the trunk road linking Rawalpindi to its new satellite many times, and memories of those journeys remain sharply etched in my mind. The chaos begins in the airport reception area, where each and every taxi driver clamours for your custom. The dilapidated taxis look quite incapable of being driven a few metres, let alone kilometres; but if they happen to be blocked by other vehicles, they will drive over pavements to get your fare, oblivious of the safety of pedestrians. Once you have been coaxed inside it is highly probable that you will find the doors and windows locked, the meter defunct, and the driver speaking little or no English. Deaf to all your pleas, he will continue to smile, assuring you he has experience of driving tanks and farm tractors. Any doubts you may have on this count are soon extinguished; so are your queries concerning which side of the road one should drive: the answer is both. As far as the driver is concerned, double white lines are only for decoration, oncoming vehicles to heighten the sense of danger, and cyclists for prangs. There are frequent detours down side lanes to outlying petrol stations, since 45-litre (10-gallon) petrol tanks seldom store more than two litres. At red traffic lights most, but not all, vehicles stop; bicyclists never. Needless to say, there are no pedestrians to be seen braving the streets of Islamabad.

The first six times I drove the dual carriageway between Islamabad and Rawalpindi I passed a dead donkey, a crash between two trucks, an over-turned taxi, a dead horse, a dying ox and two crushed bikes. Haggles over fares are frequent. Cyclists pass the windscreen from all points of the compass, swaying from left to right and back again. They wave to everyone as they pass by and may suddenly change their direction without the slightest warning. Despite all this, the experience is one to treasure and the drivers who take you through these adventures are invariably cheerful, friendly and hope you live long enough to hire them again.

Although the aerial photographs cost us several thousand rupees, they still remained the property of the Survey of Pakistan. I was also warned that these could not be taken out of the country under any circumstances, that each was marked with either a 'confidential' or a 'top secret' stamp, and that all bridges and some villages had been painted out. Our Liaison Officers were supposed to know the location of these photographs at all times, but to my mind this level of security, though strictly adhered to, was quite unnecessary and was eventually to lead to a farcical situation at the termination of the expedition.

A more serious criticism was that it severely hampered the work of our geomorphologists and surveyors. As a result, many of our scientists were forced to spend too high a proportion of their time working on these photographs in base camp, when they would have preferred to leave interpretations and comparative analyses until much later, when they were back in their home university departments; this would have allowed proportionately more time to be spent on fieldwork and the collection of data. For example, a more thorough assessment of hazards to the Karakoram Highway could have been undertaken and completed on behalf of the FWO than was possible in the short time available. But the Ministry of Defence, who were ultimately responsible for the distribution and period of loan of the photographs, stuck to their guns and so I had no choice but to take the prints on a short-term loan. Even so, I had to wait several more weeks before I could get permission to have prints made that gave us a bird's eye view of the violent landscape beyond Karimabad and up as far as Pasu.

I was soon to discover that this was not the only security issue that caused unnecessary tribulations. Both our Liaison Officers frequently had to report back to their Intelligence Organisations, and Major Rana had the daunting, if not impossible task of monitoring each and every one of our teams, who were now working at various points of the compass for unrelated and frequently unscheduled periods of time. The second officer's task seemed to me to be even worse; his job was to stay at base, monitoring all comings and goings, including the strangers that wandered into our camp.

Major Zaqir, not accustomed to the chain of command in expeditions, soon began to exercise the authority deriving from his army rank, and this led to difficulties. Eventually I had to ask for a replacement – a step I regretted, since he had natural charm of manner and we had become friends. The situation became a little clearer, however, when I realised that in Gilgit three Intelligence Officers, headed by a Captain Khan, were keeping a close watch on all our operations. Eventually I joined their counsels, and advised them to come and visit our base at any time they wished, rather than keeping in the background – a method of work that was of no assistance to anyone! This forthright approach had a sequel when I learned that Captain Khan had ordered one of the Pakistani 'scientists' home, on the grounds that he was not using the time at his disposal to the best advantage and was in effect wasting Pakistani money that could ill be afforded. Although I had tolerated this scientist, in so far as he appeared not to be doing any harm, I had realised that he was hardly what one could call a diligent researcher; nevertheless I made it clear that I expected that either I myself, or the Deputy Leaders, Nigel and Andrew, should be fully consulted before decisions were taken about members of the expedition.

This episode had an amusing aftermath. Back in Islamabad, I was informed that little attention was being paid to the mass of intelligence reports being posted back to the capital, and I was able to supply satisfactory answers to the few queries remaining in the minds of the authorities. From that moment on, liaison with the military units in Gilgit and Hunza greatly improved. We were at last above suspicion and the resulting increased co-operation brought mutual benefits.

Against this background of unease, it can be appreciated that I did not wish to cause any embarrassment to Pakistan and was concerned that expedition members should not in any way disturb this peace. We had all been screened and cleared by the Intelligence and Security Organisations, and we had all provided detailed CVs and exhibited our personal involvement in scientific research at the Islamabad conference. Accordingly, I informed all members that visitors to the Project should be discouraged. You can imagine my surprise and dismay, therefore, when one evening, while I was talking to a small group that included at least one Intelligence Officer, a young Cambridge student came to me bearing a telegram sent openly from the Post Office in Gilgit. In essence it read: 'Come to Gilgit as soon as possible, but do not mention the International Karakoram Project'. It was signed by one of our Project Deputy Directors. I could only reply to the unfortunate young man that I did not know him, could not check his credentials and that, with regret, I could not give him permission to join the Project.

Scientists who have worked many years in developing countries and who are dedicated to their work may well appreciate the conflict of interests within the structure of an expedition, but seldom do they understand the necessity of accommodating these interests, particularly in a situation such as ours. Knowing this, I had issued instructions that should a problem arise, members of the expedition should make a decision based on the risk factors involved, while understanding that if their actions led to any embarrassment to the Project as a whole, I would have no hesitation in disbanding the unit concerned in order to save the Project.

While it was transparently obvious that the young man from Cambridge was a research student wanting only to help one programme for a limited period, he had not been screened by the Pakistan government nor interviewed by me. Fortunately the predicament was solved to everyone's satisfaction when he joined his research supervisor in one of the remote villages far removed from the sensitive area. A similar situation occurred when the wife of one member came out, unannounced, late in the expedition to assist her husband's work. She too was unobtrusive and posed no serious threat to the Project. Both newcomers did valuable work, but in my opinion the risk taken was not justified.

Meanwhile the Pakistani scientists were undergoing their own severe difficulties. No money had yet been deposited in their bank account, and again they were threatening to resign. After two days of debate I had no alternative but to return to Islamabad to get to the root of their troubles. Although they had now gathered sufficient tentage, transport and scientific equipment, they had no money for petrol, drivers' wages and food. As they had all dipped heavily into their own pockets, the situation was grave. Fortunately, on the day I left to attend to this and other problems, including the collection of the final batch of aerial photographs, the first instalment of cash arrived.

After only a few days in Islamabad, Nigel Winser drove down to relieve me of what was becoming an apparently never-ending list of problems, which in totality were now taking a disproportionate amount of time to solve. I thankfully handed over these tasks to him and, after a successful state-of-the-game conference for the benefit of the international press, I left once again for base camp, clutching a microphone borrowed from the Pakistan Television Service for use by our hard-working film crew of Tony Riley and Ron Charlesworth. And now at last, I said to myself, I can get down to my professional work with the team on the Hispar Glacier, I was determined that nothing whatsoever should cause any further delay.

# The Karakoram Highway

Across the roof of the world stretches an incredible feat of civil engineering, the Karakoram Highway, or KKH for short, linking China with Pakistan. One of the many plaques along this road reads:

'Sometime in the future, when others will ply the KKH, little will they realise the amount of sweat, courage, dedication, endurance and human sacrifice that has gone into the making of this road, but as you drive along, tarry a little to say a short prayer for those silent brave men of the Pakistan army, who gave their lives to realise a dream now known as THE KARAKORAM HIGHWAY.'

Audacious in planning, monumental in construction, perpetually under repair and the cause of hundreds of deaths, the KKH is considered by many to be the eighth wonder of the world. According to one signpost near Besham, it is now possible to drive from Karachi on the Indian Ocean to Beijing, the capital of China, some 7,250km (4,500 miles) distant.

From Islamabad, the road first crosses the plains of the Punjab and then climbs the alpine passes of the North-West Frontier Province. This part of the journey alone is enough to discourage anyone from going any further. In many ways the highway is reminiscent of a European alpine road, with frequent hairpin bends, rapidly changing vistas, and steep gradients. The difference lies in the apparently suicidal tactics of the drivers who use it, the state of vehicle maintenance and the road width and surface, all of which make careful steering absolutely vital. Most passengers, who have no control of their fate, try to ignore the very real danger by closing their eyes and pretending to sleep, but it requires nerves of steel not to look up at the sound or sense of trouble ahead. Drivers aim their vehicles directly at oncoming traffic, which by definition is driven by incompetent fools. With cries of 'inshallah' (the will of God), the oncoming infidel is either forced off the road, or at least caused to take violent evasive action. Of course, the belief that God is on your side is strengthened if your vehicle is bigger and stronger than the one ahead. At the moment of truth, however, the driver with the better nervous system is he who does not care if the route he is taking goes to the gates of Heaven or the next frontier post.

Bob Holmes, our professional photographer, Helen Massil, one of the two expedition doctors, Major Rana and I left Islamabad at 8.00 am on 1 July in an army truck on the first stage of our journey up the KKH to base camp at Aliabad in Hunza. Although dilapidated, it was adequate for our purposes and in it we piled our personal baggage, alongside some of the items we had managed to rescue from the Pakistan International Airways bond warehouse. In Rawalpindi we picked up our second Liaison Officer, Major Zaqir, and his baggage, and then took the road to Abbottabad, which passes close to the site of the ancient civilisation of Taxila.

45

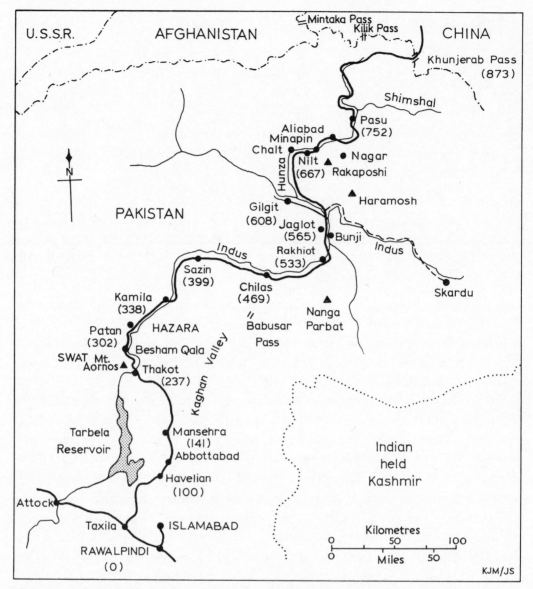

*The route of the Karakoram Highway in Pakistan, showing distances from Rawalpindi in kilometres*

It was unbearably hot inside the truck. The steel-plated sides were too hot to touch, and it was like travelling inside an oven. When we suffered our first puncture and could find no suitable jack, we stopped a lorry of similar size to our own to seek assistance. Wisely they refused to let us manipulate their jack, which had to be precariously balanced on several uneven pieces of wood, so we retreated to the shade of some trees, shooing away the lizards. Further delay was caused when we realised that one of the two spare tyres had bolt holes that did not match the truck. In retrospect, perhaps we should have kept the tyre and changed the truck.

Eventually we resumed our journey, but we had to roll down the hill into Abbottabad before stopping for lunch in order to give the driver and his mate time to diagnose the reason for statistically significant variations in the number of cylinders

46

that were firing. No doubt the state of our vehicle was due to the very heavy pounding it must have received during the long period of construction of the Karakoram Highway. A spell in the shade, cool drinks, and a vegetable type of Cornish pasty with tomato slices, probably unwashed, brought some satisfaction and time for reflection. I came to the conclusion that I would not be able to avoid picking up a dysentery-type bug at some time during this trip, so another delicious pasty disappeared in two bites. Several months later I was being treated for a stomach parasite, as were several other members of the expedition.

The truck, now rested and cooled but with the fault still unlocated, attacked the next section at 24*kph* (15*mph*). North of Abbottabad we came across many Afghan refugees, who looked in a pitiful state. Unlike the nomads who regularly move across the frontiers with their camel trains, these proud people were silent and withdrawn. Small family groups pitched their tattered tents on the grass beside the road, beneath trees that would give them additional shelter. It was estimated that by now some one million refugees had fled Afghanistan to seek shelter in Pakistan; but the main problem was the two million livestock they had brought with them which were now devastating the area, perhaps irreparably. The Pakistan government are exceedingly generous and welcoming to these people, and the local tribes have been accommodating, despite the obvious overgrazing, but the problems they are posing are horrendous and their future bleak.

We journeyed on, over and between high alpine ridges, on a road that had no straight stretches longer than about 100*m* (330*ft*). A lasting impression was made on me by the lush, bright-green patchwork of the paddyfields, embroidered into the darker green of the alpine glades and pine forests. Stack upon stack of these fields climbed the hillsides, to cascade down the far side in a similar manner. Japanese visitors to the area must be impressed by the economical use of land, which rivals their own endeavours. Our rate of progress along the road, however, was disappointing. We broke down several times, but so did several other vehicles, and I could not help but marvel at the ingenuity of the drivers, who were first-class motor mechanics using very simple tools coupled with practical ideas to effect repairs.

Some 240*km* (150 miles) after leaving Rawalpindi we descended to the Indus valley at Thakot. Soaring high above the river on the opposite bank, was Pir Sar (2,164*m*, 7,100*ft*). This formidable rock bastion, previously named Aornos by the Greeks, was captured from the Swat tribes by Alexander the Great in 327 BC as he passed on his way to conquests in India. Ptolemy, son of Lagos, was one of the generals closely associated with the capture of Mount Aornos, which was identified by the British explorer, Aurel Stein, in 1926. A plaque by the river gives details, but care should be exercised when taking directions from the bearings provided since east and west are 180 degrees out of phase!

Thakot is the spot were Alexander crossed the Indus river, and here the KKH proper starts, a macadam road of constant width, invariably described as permitting two tanks to pass each other. Judging from their antics, perhaps the drivers passing along the road see themselves manoeuvring such vehicles in the future. The road is now free of potholes, but it takes on a more frightening aspect as it travels up the

Indus Gorge. At first, however, in contrast to what was to come, the road is at river level and the only hazards appear to be the splintered boulders and fallen trees poised precariously above, all defying gravity, and the landslides of widely-varying size which block the road, sometimes once or twice per kilometre. Always there are stones on the road and occasionally boulders, too heavy for the regular platoons of maintenance engineers to push over the edge. Bulldozers regularly sweep the débris away, or if the landslide is too large for that they squash the sometimes mud-like mass as flat as possible. The Frontier Works Organisation (FWO) of the Pakistan Army is responsible for the construction and maintenance of the KKH beyond Thakot, and are understandably proud of their achievement, particularly since no private consortium was prepared even to consider the task. Despite immense difficulties they claim that no fault is permitted to close the road for more than twenty-four hours.

Between Thakot and Besham, a distance of only 28km (17.5 miles), the Indus river adds another dimension to the journey that continues almost all the way to Gilgit. The river gorge lying between the districts of Swat and Hazara has hillsides that are heavily wooded, but the lush, green paddyfields observed between Mansehra and Thakot are much less frequent, since the slopes are steeper and boulder-strewn. Also, the air is slightly cooler, because the river is now much swollen from glacier melt. The traffic, never heavy, seems more reserved now, as though the environment has chilled the fervour of the drivers. Indeed, the drivers begin to behave almost chivalrously towards one another, giving warning blasts on their horns as they twist the steering wheel first this way and then the other. Gradually I became aware that the turgid river, despite its normally quiescent flow, is immensely powerful. Cold and deep, with few points of access, it carries one of the greatest sediment loads of any river in the world ; some 5,000,000 tons per day. Nothing would survive in its muddy-brown grip and any object once below the surface would be lost from sight for ever. Here and there stretches of turbulence mark the presence of huge, sub-merged boulders, which cause spray to shoot several metres into the air. Occasional-ly several such boulders stretch across the river, probably resulting from a collapse of the cliffs above, and here the turbulence exhibits the true power of this mighty river, the lifeline of Pakistan.

At Besham, we stopped a while to repair tyres and bleed the water out of the fuel system. I took the opportunity of talking to some local children and showed them my party tricks, which include taking off the top of my left thumb and exhibiting an extra finger on my right hand. Like children anywhere in the world, they were puzzled at first, until one sharp child realised I did not have an extra finger on one hand, but a missing one on the other. As though to show their appreciation, they then attempted to sell me semi-precious stones, most of which were garnets.

Fortunately it was dark when we left Besham, refreshed by a couple of soft drinks and with the engine behaving less erratically. If it had been daylight, I doubt if any of us would have had any nerves left by the time we reached Patan, close to midnight. We could see neither the road ahead, the river below, nor the cliffs above. The dancing lights of Patan eventually shot across the darkness, only to be instantaneously blotted out and then to reappear again just as quickly, but in a totally different direction. It was an extremely eerie experience, rather like looking out from an aircraft flying through intermittent clouds. Several hundred metres above the village we descended, it seemed almost vertically, to be greeted by army officers who took us to their mess for a most welcome meal.

In 1974, Patan had suffered one of the most severe earthquakes ever recorded in Pakistan, and some several thousand people were initially reported killed. What made the disaster much worse was the inaccessibility of the area and the total lack of materials with which to effect immediate reconstruction. Our Housing and Natural Hazards research group wished to conduct studies here, and I talked to the officer in charge about this prospect.

That night we slept out in the open, beneath clear skies that were more star-studded even than the clear, clean atmospheres of Greenland. I had no desire to sleep, but was awoken from a spell of drowsiness at 4.30 am by a mess orderly with a cup of tea. Immediately after a light breakfast, I took the opportunity to walk some distance along the road before being picked up by our truck. Now the full audacity of the engineers who built the road was forcibly, irrevocably and permanently impressed on my mind. The gorge was more constricted and the road itself appeared more gravity-defying here than on any other section, but what made the situation more precarious was that the road was now approximately 600m (2,000ft) above the river. Slowly the truck weaved in and out of the innumerable gullies. Frequently, trucks coming down the valley in the opposite direction would be only 20 to 30m (65 to 100ft) away on the other side of these deeply incut clefts ; yet several minutes would elapse before we would pass, usually in the back of a gully where the road was heavily reinforced underneath with thick masonry walls. Strong nerves were required in order to pass other vehicles, and as kilometre after kilometre slipped slowly by, I found myself wondering just how long such a road could continue. All that could be seen ahead was the next precipitous wall along which the roadway clung. Evidently, this kind of scenery would continue from Besham onwards, and there was no longer any doubt as to why the tribes of the northern regions had been isolated for so long.

It was with a false sense of relief that we dropped down to river level at Kamila, 36km (22 miles) beyond Patan, refilled the petrol tanks and took the opportunity of drinking tea at a wayside shack. This respite was needed, because the worst, most horrifying section of the road lay immediately ahead. Crossing the river by the bridge built by Chinese engineers, with their symbolic lion's figures on each parapet section, we now climbed the ever-steepening cliffs of the left bank of the river. The constricted nature of the gorge was more oppressive than ever. Because the cliffs were now vertical, we lost contact with our immediate surroundings and the nearest safe ground appeared to be the opposite side of the gorge. The roadway itself was hewn out of the cliff sides, and all too frequently the cliff roof above threatened to decapitate the unwary. We were even more frightened by the splintered cliffs above our heads, which looked as if they would collapse at any moment. The river, now hundreds of metres below, could only be seen at bends in the road because the steepness of the cliffs generally hid it from view. Nevertheless, its presence was keenly felt, since trucks passing on the right forced us to the very edge of the road. On the previous section there had been perhaps a one-in-a-hundred chance of a tree or boulder halting a fall into the river, should a vehicle plunge over the edge; but here there was no chance whatsoever.

I took particular care to note the structure of the opposite walls of the gorge, since our own route would be crossing similar terrain. Before the KKH was built porters would have to descend, or ascend, hundreds of metres at each gully in order to make progress, and in many sections no route could be seen from the river's edge to the summits of the hillsides, hidden in the clouds over 1,000m (3,300ft) above. Little wonder that previously the route to Gilgit had gone north from Mansehra into the Kaghan Valley and over the Babusar Pass, although this pass was only open for a few weeks in the summer.

Along this section of the road there were no cemeteries for those killed in constructing the roadway. Anyone who lost their footing simply fell into the river, to be lost forever. On one stretch of the road a truck went over the edge filled with construction workers, and no trace of them has ever been found. This section was built by lowering men from above on ropes, who then drilled holes in which to place explosives before climbing back to safety. Both Nigel Winser (who during the course of the expedition had to drive this road nine times), and I were convinced that if any accident was to occur to our group, it would be here on the KKH.

At one point on the journey, close to Sazin, I espied a single wire rope crossing the 'Lion' river (as the Indus is sometimes called), to which was attached a bucket, deep and wide enough to transport a single person across to one of the few gullies that now exhibited any form of plant life. A few huts could be seen, and it was obvious that the best way north or south from this tiny hamlet was to cross the river by this self-propelled means and then to climb the steep cliffs to the road and wait patiently for a passing truck or bus. In this, the narrowest section, the Indus showed stretches of white water and ferocious rapids.

In contrast to the drab, dust-covered Suzuki mini-buses ferrying passengers between Gilgit and Islamabad, the lorries and buses on this stretch of the road were resplendent in exotic and colourful tattoos designed by each driver and completely covering the bodywork. The final touch of individualism was the jingling collection of fine chains attached to all parts of the vehicle. When the vehicle was in motion they would jump, trail and twine around each other as though to reflect the chaotic state of passengers and baggage inside, on top and poking through the windows. One day we saw such a bus bedecked with flowers as it drove relatives and friends to a wedding. The totally obscured vision of the driver was compensated for by his amplified horn and by the respect shown by other road users, who stopped to cheer the lurching bus on its way. Apart from these vehicles, there was little traffic beyond Patan and even less beyond Kamila.

As we left the 150km (93 mile) gorge behind, and descended to the desert valley floor, we saw no other vehicles until we reached Chilas. Here the Indus Valley is relatively wide and open, and without the protection of the gorge there is no shelter from the sun, which shines directly overhead. But now we dared to sit on the truck roof or stand on the packing cases to allow the natural draught to give some relief from the heat. Until this time I had buried myself between the spare tyres and the hot steel sides of the truck, not wishing to raise the vehicle's centre of gravity, despite the exhaust fumes swirling in from the rear and through the loose floorboards above the perforated exhaust pipe.

This was the kind of desert country I remembered beyond Skardu, and both my thirst and my smoke-filled head were forgotten as I silently thanked all concerned for getting us to this point. Helen, too, revived from her gloom and her critical but just reflections on our inability to plan even the simplest of journeys. We were without water or substantial food rations and subjected to temperatures around 55°C (131°F), and we would have to wait until Jaglot before we could slake our parched throats. Even our two Liaison Officers, both of whom had had tours of duty on the KKH, were not sure where we could find supplies.

We passed the checkpoint at Chilas without even noticing it and understandably the officer in charge was very annoyed at having to chase us upstream in order to gain our signatures for his record book. Every vehicle that contains foreigners must be stopped at each checkpoint, and registration number, permit number, date and time must be entered in the book before permission can be granted to leave for the next stretch of road. The checkpoints are usually situated among wayside huts and consist of a slender tree branch stretched across the road at about chest height. This branch is pivoted at one end with counter-weights of small concrete slabs with the thinnest end of the branch slotted into a groove on a small post at the far side of the road. In order to allow villagers to pass to and fro, the branch is usually in the vertical position, and since pedestrians are constantly dashing across the road, blocking the view of attendant policemen who, if not drowned in the dust cloud, are unrecognisable in the crowd, it is all too easy to pass through unwittingly. To complicate matters, the checkpoint officers make no attempt to stop oncoming vehicles if only Pakistanis are aboard, and during latter stages of the expedition most of the expedition members were almost indistinguishable from the local inhabitants in both colour and attire. On subsequent journeys all the checkpoint officers were inquisitive but pleased to see the Land Rovers displaying the three flags of China, Pakistan and Great Britain. Indeed, once they came to recognise the Land Rovers and learn of our objectives, strong friendships were established.

Our four Land Rovers were generally accepted as ideal vehicles for this kind of country. Bob Stoodley had done a first-class job in preparing the vehicles and his many hours of effort and financial outlay to put them in condition to meet any difficulty was one of the great successes of the project. The V8 engines gave more than sufficient power and the versatile third gear coped admirably with the innumerable bends on the KKH. On one journey Nigel travelled all the way from base camp to Islamabad, a distance of 720km (450 miles), in a little over thirteen hours.

The section of the KKH we were now travelling on requires little maintenance, since it is built across a plain that suffers little erosion. This plain is also of

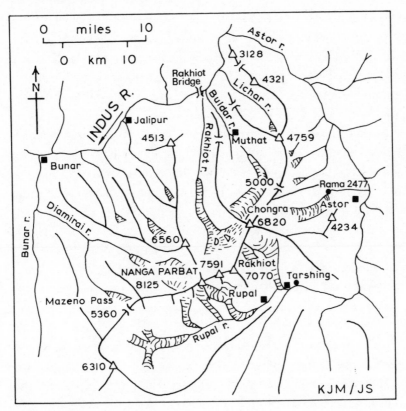

*Nanga Parbat and environs*

considerable geological interest. Large boulders are dotted around, covered in a thick, brown shell called desert varnish and with deeply-eroded pockets that may be water-worn. Probably the only form of violent natural action in this sweltering open valley would occur when the transient dams on the Indus or its tributaries collapse and 15 to 24m (50 to 80ft) high walls of water flash down the gorges to devastate the Punjab. Could these floods or glaciers have brought down these boulders, or were they a product of wind erosion and bombardment by sand particles?

The river had once again reverted to its deceptively tranquil pace, but occasionally a turquoise stream descending from the high snowfields and alpine pastures above would emerge from a side valley, twist around as though to accommodate the major direction of flow, but be forced to yield its strikingly beautiful colour and sparkle to the brown opacity of the uncompromising Indus. Now more and more snow-capped peaks began to emerge, and glaciers could be seen threading their way down into the higher valleys. At Rakhiot we crossed the Indus for only the third but last time, and soon were heading for one of the major bends in the river as it forces its way through the countless folds of the Himalaya and Karakoram and the hills of the bounding states of Hazara and Swat. Just before Jaglot, we stopped and looked at the snow and ice mountains to the north, among which stood Rakaposhi and Haramosh. Turning to the south, we saw the lower cliffs of Nanga Parbat, the summit being hidden in thick cloud (see above). I doubt if there is any other place on earth where it is possible to see three giants such as these. Their great precipices fall more than

The Pasu group of peaks opposite the Batura Glacier.

Aliabad in Hunza with Rakaposhi in the distance.

*Top* The Karakoram Highway below Pasu;
*above* Porters negotiating a disintegrating section
of the track to Hispar; *left* Land Rover returning
to base from Nagar – note cascading fields in the
background; *over* Part of the Hunza gorge – note
the scree slope on the right.

6,100*m* (20,000*ft*) from the summit, through snowfields, fluted ice walls, crevassed glaciers, roaring melt streams, avalanche-prone rock buttresses, high pasture land, alpine glades, hamlets, villages, desert-like plains, and finally to the river that carries away all the débris that tumbles from on high.

As I looked across the river at Bunji, I remembered the daring and arrogance of Hayward and the apprehension of Vigne so long ago (see Appendix I), and it was while reflecting on the efforts and tribulations of these explorers that I was brought back to earth by yet another exasperated checkpoint officer, calling us to return many kilometres downstream to fill in his book. What purpose these books serve, I do not know, especially since we had permits that would allow us to fly directly to Gilgit from Islamabad.

At Jaglot we pleaded ignorance of his post, showed him our permits, mumbled about international co-operation, drank tea, went to sleep, awoke and started pleading again. Radio calls went north to Gilgit and south to Chilas. Eventually all was settled by my writing a letter stating that he was not to blame for our omissions, that we would rectify our mistake in Gilgit, and finally that we were sorry to have caused so many problems to this efficient and friendly officer. All of which took twenty-seven cups of tea before we could continue.

My one regret on this first journey was that somehow I missed the confluence of the Indus and Gilgit rivers. Perhaps this was because the gorge through which the Indus comes from the direction of Skardu has a very narrow neck ; a tortuous valley that hides its flanks in the uniform colour of the giant hillsides. Perhaps, also, it was because the confluence is well below road-level. Or perhaps my eagerness to look ahead and spot the first signs of the intriguing town of Gilgit simply caused me to miss this junction where the Indus escapes the grasp of the Himalaya.

Gilgit is built on a huge alluvial fan which flows out from the mountains to the south of it. The entrance to the town was guarded by yet another post, attended by an alert officer who repeated the procedures of the other eight checkpoints. Here, the roadway was washed away in two places, but was still passable; the damage would be repaired in the winter months when there is greatly reduced flow in the rivers. We went straight to the Chinar Hotel, met Nigel Winser, washed, unpacked and brought ourselves up to date with plans for future operations.

Early the next day I forsook the truck and journeyed up the Hunza Gorge in one of the expedition Land Rovers. Crossing the impressive Chinese bridge over the Gilgit river, we passed through Daimyor and our final checkpoint before base camp. The thrill of entering the greatest of all the Karakoram gorges cannot be expressed in words. It simply has to be seen, felt and absorbed. From Gilgit it was difficult to distinguish where the Hunza river outlet was located. The entry up the grey, daunting, vice-like cleft looked highly improbable. I was, and still am, amazed by the efforts expended by competing armies that have sought to control this valley in the past. The 100*km* (62 mile) journey took us only four hours, but nevertheless I still cannot understand how any pre-KKH traveller journeyed to Hunza in less than four days on foot or horse. The journey from Islamabad to Gilgit had not involved any great gain in altitude, but now we were to climb some 1,200*m* (4,000*ft*), through a gorge that had defied many invaders. The valley walls are steep and fragile and the slightest rainfall brings down great slides of débris that frequently block the road.

Close to Chalt we entered the zone where it is thought the continental plates have collided, and where the gigantic forces involved have created the most chaotic landscape on earth. The Land Rover sped quickly under great overhangs of split rocks, and the sight of huge boulders in the fields below the road was witness to the ever-present danger overhead. Long before Nilt, we stopped after rounding one slight bend, because there, suddenly unveiled before us, stood the proud mass of Rakaposhi. This mountain was to be our base camp backcloth for three months, and never once did it fail to impress and inspire. In the morning and evening it showed different colours and shadows; sometimes it was veiled, sometimes clear. Only when its slopes were carefully inspected with a climber's eyes did it become intimidating. (A brief history of the ascent of Rakaposhi is given in Appendix V, p. 203.)

The advance of the British army up the Hunza Gorge in 1891 had been checked at Nilt, and only the bravery and tenacity of a few shock troops who climbed vertical crumbly cliffs hundreds of metres in height overcame the stiff resistance of the defending Nagaris. Beyond Nilt and close to Hini an old track was pointed out to us where several jeeps had lost control on the loosely packed hillside and had had an almost clear fall into the river below.

The distance between villages here varied between 10 and 20km (6 to 12 miles), but they were always set against a backcloth of steep hillsides. One half of a village had slipped more than 30m (100ft) during a landslide, but its fields remained intact and life continued as before, except that the climb home at the end of a day's toil in the lower fields was now that much longer. The major features of the Hunza Valley, however, are the great glaciers that plunge down the clefts in the sides of the valley and threaten the villages below. It is these glaciers that cause the greatest problems to the KKH. Sometimes they advance rapidly and not only completely sever the road, but also dam the valley. At all times they send down vast volumes of melt-water and rock that cause serious erosion, and both glaciers and melt-water deposit large quantities of débris. The Hasanabad Glacier, for example, has advanced by almost 5km (3 miles) in the period between 1954 and 1980, and is known to suffer equally rapid retreats. At the turn of the century the glacier advanced by approximately 9km (6 miles) in one season. As a result, watercourses can change direction spectacularly, and hence the planning of bridges to carry the KKH is exceedingly difficult. At zones where it is impossible to tunnel under the roadway or build bridges over the watercourse, several parts of the road itself act as stream beds.

Foreigners are not allowed past the Chinese bridge at Ganesh, beyond Aliabad, but by special dispensation we were permitted to go to Pasu (see p. 100). It was this section that proved of most interest for our scientists, since it is here that the most destructive forces of nature — avalanches of ice, snow, mud and water — have been at work. Just beyond the bridge, frequent rockfalls occur. I was always pleased to get beyond this stretch and be able to relax on the short but relatively open piece of road just before the gorge at Sarat. It was here in the gorge, in 1857, that a great slab of rock and glacial débris collapsed into the Hunza river; over the next few months the water level rose behind the dam until, at the height of the summer flood, the dam disintegrated and a surge of water sped down the Hunza, Gilgit and Indus rivers, the latter rising by 17m (55ft) within a few hours, before losing its mass on the plains of the Punjab at Attock, some 700km (440 miles) distant and in the process drowning an entire Punjabi army.

Today the remnants of the dam are still to be observed as the river charges through a boulder-strewn ravine in which no human could survive. Meanwhile, the cliffs above the dam site are still rotten and threaten to deposit hundreds of thousands of tons of rock into the ravine and cause another catastrophe of equal dimensions. Close to this point, a crushed vehicle could be seen below the roadway. Perhaps some of its occupants had escaped, since this was one of the very few sections of the entire road that did not fall directly to a watery grave.

The road frequently passes under a long line of rock cliffs in which are embedded huge, shattered, angular blocks. The orientation of the cracks in these blocks is such as to facilitate the disintegration of the cliff, and it seemed to us that the road might be safer if more bridges were built, allowing it to alternate frequently from side to side. However, it was clear that repairs were effected very rapidly, and of course, a roadway is far less expensive to replace than a bridge.

At Shishkat, which lay around the next corner, a mud-slide in 1974 demolished half the village and also blocked the Hunza river. The resulting dam created a lake some 15km (9 miles) long, that drowned both the road and the bridge, the latter now being lost under a few metres of new river-bed silt. The FWO took this particular challenge in its stride and within seventeen hours opened a temporary stretch of link road along the near side of the valley – although, of course, the construction of a new bridge and a new road of comparable standard to the now-drowned portion took more than two years to complete. En route to Pasu, we crossed over the magnificent new Chinese-built bridge and weaved among several gigantic boulders which had fallen recently and only just missed crushing the now uninhabited checkpoint building. The maintenance engineers in this section were now focusing all their efforts on the problem at Gulmit, where the Ghulkin Glacier melt-stream had swept the highway straight into the Hunza river. The Land Rover drove majestically on over the rubble-strewn river bed, to the envy of other drivers halted here who were waiting to be towed across by army engineers. Then we came to our journey's end at Pasu, where we camped in the old huts of the now-departed road-builders.

From this point we visited the Batura Glacier snout. Despite our knowledge of the multifarious forms of disasters with which the FWO have to cope, we were stunned by the sight of the remnants of the two-lane concrete bridge, now lying in the river bed. A glacier melt-water surge had simply crushed it like matchwood. The new bridge was of a temporary construction, but adequate to carry heavy two-way traffic. Beyond this point we dared not venture as, regretfully, we had not yet received permission from the appropriate authority, but the Pakistan and Chinese members drove on up to the border and informed us that the remaining part of the KKH was a straightforward drive presenting no special difficulties.

In subsequent statements to the Pakistan press, I reported that in my opinion the KKH was a modern wonder of civil engineering and should be viewed as a symbol of national unity. It not only provides the most valuable link in the north-to-south road system of Pakistan, but joins two great countries together. Little wonder that it is called the 'Friendship Highway', a name that will be a permanent and fitting memorial to all those Chinese and Pakistani workers who died building this road across the roof of the world.

The KKH served our expedition well, permitting rapid movement between, and deployment of, different research groups. No doubt it will facilitate future scientific studies too, particularly in geological research into plate tectonics (see *Earthquakes*, p. 86, for a full explanation). It is possible to study many outstanding problems concerned with the origin of continents simply by driving along the KKH.

The preceding description of a journey along the Highway is, of course, based on my own personal experiences, and so to complete the record it is worth recounting the story of another member of the expedition, Roger Bilham, one of our geophysicists, who travelled the KKH back to Islamabad on his way home. If the accidents he recounts had been in reverse order, he would never have been able to tell the tale:

'It was my misfortune to plan to travel from Gilgit the following day, a Friday, and to arrive at a time when all the buses and planes were already booked. It was thus a relief to be informed by the hotel manager (who asked me simultaneously how much I wanted for my shoes) that a senior political figure would be travelling by car to Gilgit early the next morning and would value my company on the long drive to Islamabad. My transport problem solved, I determined to heed his advice and be ready at 4.00 am, so as not to miss the air-conditioned Mercedes parked promisingly near the hotel entrance. 3.30 am found me shivering slightly beneath the stars. 5.30 am found me anxiously watching two tired-looking workers loading up a decrepit Suzuki jeep. At 6.00 am its owner, a diminutive businessman who could have been a grocer or a bank manager, clambered aboard and beckoned me. The hotel manager had spoken of a famous scientist wishing to travel to Islamabad. Did I know of his whereabouts? I climbed in.

'Thirty minutes later, a loud bang beneath the piles of plastic cups, cheap reproductions and Chinese radios announced that all was not well with the top gear, and that a more moderate choice of gears would be advisable for the next seventeen hours. The driver shook his head and hinted at twenty-five hours. The businessman looked grim. Meanwhile, a tickling sensation about my knees revealed that, in the absence of a cover, battery acid was etching its way through my jeans. This might have been annoying in other circumstances, but in this instance the engine cover was also missing and the blotchy holes provided much-needed ventilation for the jet stream from the over-worked engine.

'Hours passed as we made our way along the Indus Valley, with stupendous drops on one side of the road and towering cliffs on the other. My mind was occupied by the views, the challenge faced by the road construction engineers, the remarkable terraced farming, the overloaded trucks, the apparent contradiction between the piles of shoddy merchandise in the jeep and the exercise of accountancy, which my host was relating to me in some detail but without enthusiasm.

'Suddenly we were approaching the back of a highly-decorated but stationary truck. Instead of stopping, the jeep glided rapidly on between the truck and the vertical cliff and fortunately fetched up on a large pile of sand. The driver, my host and I walked around the jeep a few times and pushed it out. The verdict was a loss of hydraulic fluid. Since there was none to be had nor anyone around to help, the driver decided that we would drive the 32*km* (20 miles) to the next village without brakes.

'The views now became of secondary interest as we edged our way along relaxing straight bits of road and fairly tense curved bits of road and rather pessimistic-looking downhill and uphill bits of road. The sumac trees bordering the precipice began to look supremely flimsy and the occasional stone wall became a reminder that even the engineers thought that some parts of the road were more dangerous than others.

'A new sound indicated that entropy was at work again. We came to a slow, graceful halt to find smoke coming from what used to be the rear tyre. The spare wheel was introduced and inflated at the next village with a pump. An engineer fixed the hydraulic brakes over lunch, and five hours later, one Chinese radio less and one plastic tea set more, we were on our way.

'Darkness forced us into a hotel, because earlier that day the headlights had gone the same way as the brake fluid. My jeans had by now bifurcated at the knees, and since we had left the Indus Gorge and reached the humid lowlands, I detached the lower parts to increase ventilation and provide a kind of plug for the battery acid. The evening was passed in the pleasant company of a Pakistani English professor, who discussed the Leeds dialect versus received-pronunciation from the point of view of Polish immigrants in the UK.

'Since I was to travel from Karachi the next afternoon and was still somewhere in the Karakoram foothills, I was anxious for an early start and not too many mishaps. The post-breakfast surprise was a new grinding noise from the back wheel and the simultaneous loss of more brake fluid. The driver cooled the wheel down by converting several buckets of water from a nearby mudhole into superheated steam. His verdict was that we would probably get to Islamabad if we took it easy. About five minutes later an extraordinary change in the character of our motion revealed that we were now travelling on three wheels and, sure enough, visible through the side window was an independent wheel, bumping merrily over the fields, complete with half shaft and a sprinkling of gears. A shower of sparks, pieces of brake pads, slave cylinders, pipes and assorted metal fragments mapped our slow, drunken deceleration to the base of a gentle hill. An interesting ride.'

# Illness, Accidents and Tragedy

One of the dilemmas faced by those organising an expedition is how to safeguard its members against unforeseen circumstances. On previous expeditions for which I was responsible, I have always declined the opportunity to include members of the medical profession, and on most occasions I have also not taken radio sets that could be used to call for urgent medical aid. This was not out of bravado, nor to simulate the conditions of the early polar expeditions, but for pragmatic reasons. Radio sets are heavy and cumbersome, and militate against fast-moving, lightweight mountaineering ventures on which members have to carry everything they need on their own backs. Doctors, on the other hand, can provide a distraction that is hard to resist if the weather is poor, the body tired, and when thoughts of home comforts and meals erode the desire to get things done. Malingering seldom occurs on an expedition, but it can happen, especially when danger lurks around the corner.

Experience has taught me that when there are no doctors and no radios you have to get on with the task in hand and rely totally on your own abilities to get out of trouble. It makes you very aware of your companions' needs and generates self-confidence and the extra care which is necessary to avoid accidents. To compensate for the lack of medical expertise, however, there should be at least one member of the team who has completed a first-aid course, and every member should carry his share of a medical pack, the contents of which should be selected on the advice of the medical profession. Under these circumstances, it is very satisfying when the team returns, unscathed, to its base. This, more than any other physical or mental achievement is the one that permeates the whole soul during those few days it takes for members to return from base camp to the warmth and comfort of civilisation.

With such entrenched attitudes, it was very difficult for me to reverse my position, but I had no alternative. There were seventy members of the Karakoram expedition, all working across difficult terrain. Without doubt all of us would catch some form of stomach bug ; and in a land where few if any medical resources were available and where some groups might be separated by 100km (62 miles) or more, it was essential to have a doctor among us who could both reduce the number of lost days' work to a minimum, and keep in regular touch with all groups by radio to monitor their state of health. At no time, however, did I seriously consider that the statistics relating to accident-probability were weighted against us – although with more than 5,000 man-days of involvement, there was more than a slight chance that one or two accidents would occur.

Helen Massil, together with five other applicants, had previously written to the RGS, asking if any expedition required a doctor. Recently qualified from Sheffield, she had come along to my office for a talk. Her enthusiasm was obvious, and despite her lack of expedition experience, an important point in her favour was that a woman doctor would be able to gain access into Muslim homes. On previous trips I had noted that only the male population had come to our tents to seek assistance during our evening halts, and only on a few occasions did a father bring his sick baby daughter for treatment. We were never invited to their homes, nor asked to assist in the medical care of their womenfolk, many of whom were either afraid of foreigners or were not permitted to have treatment from them. On the other hand, the Muslim community obviously cared for their families ; by including a woman doctor in our group, we therefore knew that we could achieve a far more rapid rapport with the villagers.

David Giles, a much experienced doctor with a large general practice in Bude, Cornwall, was selected as the senior Expedition Doctor. He brought home from the Mulu expedition a reputation that was second to none – so much so, in fact, that I was a little sceptical, especially since I had previously not appreciated the necessity of including a doctor at all, let alone two.

How wrong I proved to be ! David and Helen were pillars of strength and provided a bonhomie that affected us all. Sympathetic but matter-of-fact, they combined care and attention with the desire to effect a speedy recovery ; and without their attention we would have lost many man-days of effort and would never have achieved the support of the local population to the extent we did. Theirs is a success story in its own right.

Medical care began before leaving the UK, when all members of the expedition were advised of necessary and optional innoculations and on general measures to improve health while in the field. The medical team also took the opportunity to communicate personally with as many members of the team as possible, both by letter and at the pre-expedition get-togethers at the RGS. Supplies of both drugs and medical instruments were in hand long before departure – the British pharmaceutical industry responded magnificently to our appeals by supplying the expedition with approximately £5,000 worth of drugs. The provision of drugs was designed to meet both the immediate expedition needs, and also to provide medication, etc. for the local people, whom we expected would request medical assistance from us. It was expedition policy to look after all our own medical problems as far as possible, and also to treat local people for acute illnesses. Clearly we could not institute treatment for long-standing organic problems, as this would have to be discontinued when the expedition left the area.

For the current project, all expedition members were provided with their own field medical pack, which was found to be very useful. In retrospect, although a few items in this pack were under-provided and others over-provided, on balance it was effective. Additionally, each project leader was issued with replacement drugs and dressings so that members' packs could be kept topped up in the field.

The initial choice for our base headquarters and, as it turned out, our permanent base, was a field at Aliabad given over to the Pakistan tourist board for the summer season. On arrival at this camp site, priority was given to the provision of a safe water supply and adequate washing and cooking facilities. We were fortunate in finding clear water close to the base camp in Aliabad, and alternative supplies were discovered subsequently, although further away, in some cases up to 20 to 30*km* (12 to 18 miles) distant. These sources were used when the Land Rovers were taking or fetching working parties. The provision of a plentiful supply of clear, safe water was a considerable bonus to our cooks, who were extremely wasteful in its use. Some water was also used from the standard camp supply line, but only after it had passed through our filter and chlorination system. It was subsequently discovered that some of our water came from springs at Hasanabad that were heavily laden with magnesium sulphate, and this was thought to have caused some of the stomach upsets. The initial, unintentional consumption of large doses of such salts was curtailed towards the end of the expedition by the use of other supplies from further afield, but not before we had all succumbed to Delhi Belly or the Karachi Quickstep. These debilitating illnesses may also be caused by local bugs, but the effect and treatment was identical.

It proved nearly impossible to educate the domestic staff at the camp in matters of hygiene, and despite entreaties, demonstrations, bullying and other tactics, we failed to encourage them to act in a hygienic manner. The worst threat to health came from the numerous flies. At night these would black out all forms of lighting, and during the day they would blanket any available food, whether it was moving towards someone's mouth, inside containers waiting to be served, or on minute crumbs on a beard or someone's lips. Eventually, with the use of flyscreens we managed to keep some of the worst of the fly contamination out of the kitchen, but we still failed to convince our cooks of the importance of washing their hands before and after the preparation of food.

David Giles supervised the maintenance of the toilets, which proved to be a mammoth task. Numerous breakdowns occurred, some caused by the postures adopted by visitors who were not accustomed to the flush system and toilet pedestals. It was only with difficulty that people were stopped from standing on the toilet pans in order to empty their bowels, and it was a hopeless task to persuade the camp employees to undertake adequate and regular cleaning of the toilet and shower block. Possibly the conditions in this block contributed to the gastro-enterostinal problems seen in the camp, especially since the camp staff seldom washed their hands. Nevertheless, it has to be said that, considering the density of the camp population, it was remarkable that no full-scale gastro-enteritis epidemic occurred. This was entirely due to the efforts of David, who provided separate clean water for cleaning teeth and did his utmost to change the habits of the camp staff.

During the course of the expedition, medical records were kept for all our 73 members and visitors. These record some 260 full consultations with Dr Massil or Dr Giles. Of these consultations, about half were for diarrhoea or diarrhoea plus vomiting, most commonly caused by bacterial enteritis. Respiratory tract infections produced a small but significant number of problems, but the trauma was remarkable by its infrequency, particularly considering the terrain in which work was being daily carried out.

Seven cases of hepatitis occurred, which appeared to be derived from an original infection at base camp. Three of these cases required evacuation to the UK, simply because it was unlikely that full physical recovery would have occurred before the end of the expedition; to have allowed those affected to remain would have provided a possible source of infection, while achieving no scientific objectives. The remaining cases occurred late in the expedition, while two members did not exhibit the full symptoms until their return to the UK. Of the seven stricken members, none had opted for protection by the use of human antihepatitis gammaglobulin. However, even if they had the potency of the gammaglobulin would have been reduced by the time that they caught the infection.

The first to succumb was George Musil. Perhaps a feature of this debilitating disease is that it strikes the toughest far worse than anyone else. George had already proved that he had more energy and more lasting power than most; the size of his muscles may have been slightly exaggerated, but no-one would have chosen to wrestle with him unless drunk. Now his appearance gave the impression that he had tasted one bottle of Hunza water (a lethal and illegal brew) too many. He tried his best to fight off the lethargy, and after a day's rest he would insist that his recovery was now complete. However, it took no more than a few minutes' walk around the camp before he was fatigued once more. Clearly the infectious disease spelt danger to the whole project, and David Giles quite correctly ordered poor George home. He was heartbroken, but there was no alternative.

The other six sufferers – Tony Riley, Paul Nunn, Ron Charlesworth (all members of the film unit), Nigel Winser, Jon Walton and Shane Wesley-Smith – all had enforced periods of rest in UK hospitals after their return. Fortunately the strain of the virus that they had contracted was not the most dangerous type, but it certainly is to be avoided if at all possible. All those affected bitterly bemoaned the fact that they could not touch alcohol for six months after they left hospital.

Bob Stoodley, too, had a spell in hospital in between his two visits to Hunza because of a parasite, and the take-care telegram he sent us from his bed, telling us that the Hunza Hop was the least of our worries, caused much amusement as well as sympathy for his predicament.

The medical work amongst the Hunzakuts was fascinating and rewarding. There had been no resident doctor in the valley for twelve years, although doctors had visited the area for short periods of time. The small hospital at Aliabad was very run-down and was operated by one dispenser, Hamayun Beg, and two or three willing but extremely ill-educated assistants. The supply of drugs to this dispensary was haphazard in the extreme, and the quality of the drugs was not always appropriate to treat the local endemic diseases that occurred. Hamayun Beg gave great assistance by acting as a link-man between the local population and our medical team, and his extreme kindness and attention to duty overcame the great stress and grossly inadequate facilities with which he normally had to contend. It is, perhaps, unfortunate that his service to the Hunza community receives little acknowledgement.

Dr Massil and Dr Giles spent between one and three hours in the dispensary every day, with the exception of the Muslim and Christian sabbaths. They were also frequently called to visit the houses of the local population and usually examined ten or twelve of the principal sufferer's family in addition. Several fascinating and extraordinary cases and methods of treatment were observed, many of which would never be seen in the UK. For example, open wounds would be sealed off by packs of mud baked dry in the heat of the sun, with no account taken of the constituents of the water, silt or clay that comprised the cast. Surprisingly, the doctors found many anxiety- and stress-related illnesses, and also hypertension and heart disease, though some of the disease may be attributable to the rapid social and economic changes forced upon the area by the construction of the Karakoram Highway. Not so surprising were severe and frequent gastro-intestinal problems, notably the occurrence of peptic ulcers. These ulcers were almost the rule in males between the ages of 20 and 40 and occurred very frequently in women between 30 and 40. It was also noted that gastro-enteritis was much more common in Aliabad than in Karimabad (see map, p. 100), apparently because the water supply became increasingly contaminated as it came down the mountainside. The drinking water also served as the irrigation water, flowing from ditch to ditch across fields that had previously been fertilised with both animal and human excreta. There was a significant amount of congenital mental disease and also a number of other congenital diseases such as dwarfism and albinism. There is no doubt that the incidence of all these congenital problems was infinitely higher than would have been seen in the UK. Thyroid disease was expected and was seen. However, it was surprising that the majority of thyroid complaints seemed to occur in women and not in men.

Members of the medical team were fortunate in being able to visit various parts of the region outside the Hunza Valley, and they found that the pattern of illness and lack of medical facilities was repeated in other places such as Hispar and Yasin. The visits outside Hunza provided a unique opportunity to see the life of the various people of the Karakoram in and around their villages and in their homes. We were all impressed with their friendliness and generosity, but were appalled at their ignorance of ordinary hygiene and baby care. There is effectively no contraception, largely on religious grounds, although we felt that, with a little persuasion and better use of the Health Guard System which has already been introduced in some villages after efforts by the Aga Khan, some form of contraception *could* be introduced. Certainly there is urgent need for medical education and the provision of a safe, clean water supply. This latter would be perfectly possible by taking water from the existing streams before they cross the fields, filtering it to remove the majority of the suspended silt etc., and then piping the supply to various convenient points. We estimate that this would reduce the morbidity amongst the local population by at least 80 per cent. Further, we feel it is vital that good medical supervision is provided for the valley, which is presently grossly under-doctored. The provision of adequate and correct supplies of drugs could be achieved relatively easily without greatly increasing costs. We believe that the people of Hunza and of the Karakoram areas in general are not receiving the basic medical care which is their due, although the Aga Khan and his organisation are doing much to provide paramedical and social care in the Hunza district.

The doctors were called on day and night, and visitors came all the way up from Gilgit and sometimes from further afield to attend the daily clinics. Malingerers were turned away and drugs only handed out to the obviously sick. The numerous pleas for drugs to ease the pain of relatives too ill or too far away to come to the clinic had to be firmly but politely refused. From treating the local population, we quickly became aware of the health hazards to which expedition members were subjected, especially because their resistance to the indigenous strains of diseases would be low. During the course of the expedition the wisdom of having two doctors with us became more evident to me. Mountaineering expeditions to the Karakoram, unlike our own, usually speed through the valleys and within a matter of a few days have set up a base camp on a glacier below a mountain that provides a plentiful source of clear, clean water, far removed from the villages of the valleys. Furthermore, it is much easier to convince your porters on these expeditions of the need to regularly wash and to ensure that they do. I recalled that on my previous expeditions here sickness was quickly cured once we had established base and started our climbing.

The first serious illness among members of the expedition occurred before we started. Bob Stoodley, Jon Walton, Nigel Winser and Steven Redhead had just collected the four Land Rovers from Karachi and because they had been delayed by customs officials and now wanted to lose as little time as possible before setting up a base camp, each drove a vehicle non-stop from Karachi to Islamabad, a distance of 1,580km (almost 1,000 miles). During the heat of the day, when outside temperatures in the Sind Desert were in excess of 54°C (130°F), the heat generated in the cabs above the engines caused temperatures to rise to approximately 65°C (150°F), sufficient to melt pieces of chalk on the dashboard, leaving stains which could not later be removed. On arrival in Islamabad, both Jon and Steve were dehydrated and extremely exhausted and had to be put to bed. They both consumed large quantities of liquid, and it was soon apparent that, while Jon would recover, Steve was very poorly indeed. No sooner had Helen Massil arrived by jet from the UK than we had to ask her to return to London immediately with Steve as a patient. Faced with this unenviable task she displayed an equanimity that was to mark her future activities, and despite the possibility of a double dose of jet-lag, made all arrangements quickly and efficiently. On arrival in London she got Steve, still in an exhausted state, into hospital and contacted his parents before turning around and flying back to Pakistan.

The next serious case concerned Professor Zhang, our glaciologist and the leader of the Chinese team. While crossing the Hispar Glacier, he slipped, apparently spraining his ankle, and had to be carried to his tent, which was situated close to moraines in the middle of the ice flow. There he decided to rest his foot and await improvements.

Simple accidents such as this happen to the most experienced climbers : a loose stone, only partially embedded into the glacier surface ; a piece of ice that has been in shadow and so has a surface that is smooth and slippery, rather than sparkling with hundreds of thousands of water-filled micropores sufficient to give a grip ; a friable piece of ice ; a nudge from the other boot ; a foot placed carelessly, or any other normally inconsequential event – all can cause that instant of instability that can lead to a fall. But here in the middle of a glacier, several days' march away from the nearest village and at least seven days from Hunza, such an accident can have far-reaching results. Zhang did his best to convince himself and his companions that his sprain would soon mend, but his hobbling became more painful and we began to suspect a fracture. David was soon due to return to his practice in Cornwall, but he made a dash to glacier camp from the Hispar base camp, where he had recently pronounced that George Musil, our first hepatitis case, should be evacuated. Sadly, he too suspected an awkward fracture and so radioed back to Hispar base to get us to ask Hunza for a helicopter on the next scheduled radio call. The safe removal of Professor Zhang to Aliabad took two days to organise, owing to bad weather conditions, but he stoically remained calm and insisted he would soon be back at work. Alas, Zhang found the plaster of Paris cast uncomfortable and difficult to manage and so returned to China, since it was obvious he would be out of action for some considerable time. (See *Into the Ice*, p. 127.)

Before Zhang's accident, Timothy Moughtin, who was helping his father Cliff on the Housing programme, had had to be brought down from the high valleys. He was reported to be eating nothing at all, and if coaxed to taste a little solid or liquid food was immediately sick. With no doctors available on their site, Cliff wisely brought Tim down to Gilgit, where both doctors examined him, but could find no clue as to his malady. Because Tim had had little or no food for almost two weeks, Cliff had no choice but to abandon his work and take his son back to the UK. I met them in Islamabad, primarily to see if we could get Tim back to the UK in the company of a British nurse, so permitting his father to return to the expedition, but on seeing Tim I knew this to be unwise. I have never seen a boy so pale and wasted; it was remarkable that he could still smile and accept my cheerfully rude remarks about the lackadaisical state of modern British youth. It was not until he was in hospital in the UK that he was found to have a parasite in his stomach.

On my previous trips to the Karakoram, I too had picked up parasites, but it was not until I had a check-up in January 1981 that my 1980 variety was identified. This time, however, instead of being painfully cleaned out by medical instruments, I was given small yellow pills to be swallowed with water. I, too, had to be abstemious.

Ron Ferrari also had to go to hospital on his return to the UK. He had slipped on the walk back from Hispar to Hunza and, although suffering a badly grazed shin, he was not significantly troubled until he arrived home approximately ten days later, when the wound turned sour and took far longer to heal than anticipated by the doctors.

Accidents and illnesses such as these could not be foreseen. However, statistics did point to an ever-present danger about which I was particularly apprehensive, namely, the threat to expedition members posed by the perpetual disintegration of the mountain sides. Andrew Goudie, David Jones and Professor Tahirkheli all had near escapes from falling rock in the Hunza Valley. Eventually Alan Colvill, a member of one of our survey teams, was also involved in such an accident – not altogether surprising in view of the fact that the teams were constantly climbing

3,000m (10,000ft) hillsides in order to visit one survey station after another. Alan was crossing a dangerous stone chute when he heard a cry to beware. Unwittingly and spontaneously he looked up as a flying slate arrow whipped past his face, leaving a nasty gash that required several stitches just below the eye. If he had been taller, or moving faster, or had set off sooner . . . But then, this kind of analysis is never constructive and could be applied to all our mishaps.

The most serious potential accidents were those concerned with journeys between villages. The chapter on the Karakoram Highway highlights the kind of dangers that exist, but these are magnified on journeys on the narrow mountain tracks, over which only tiny jeeps can manoeuvre. Here the hairpin bends are so acute and ridiculously narrow that the upper part of the bend frequently collapses on to the lower part and the track has therefore to be constantly rebuilt. At one point in the Hunza Valley, close to the village of Hini, a track is so unstable that no less than four jeeps have toppled over the edge and plunged down into the depths of the Hunza river. In Gilgit the UN peace-keeping forces informed us that the first modification they insist on when receiving replacement vehicles is to have the doors removed. At the last count their group had lost five vehicles, four drivers and no UN observers. These stories are told factually and are not exaggerated. If we were to have a serious accident, then in all probability it would be associated with the jeep tracks. Graham Yielding, a member of the seismic group, tells the following story:

'James Jackson, myself, Rehman Khattack and our Pakistani driver set out early in the morning in our jeep to drive to Babusar, close to the pass of that name that links the Indus to the Kaghan Valley, loaded up with seismograph, seismometer, batteries, etc. to establish a seismic station there. The jeep-track follows the wanderings of the valley, and we could average no more than 16kph (10mph). We rapidly decided that, rather than do this journey every other day (to change the record on the seismograph), it would be better to locate our station near to the Babusar rest-house and for one of us to stay there to maintain the instrument.

'After two and a half hours' driving, which took us 35km (22 miles), we reached Babusar village, an odd collection of dark log cabins, which looked as if they belonged to the Wild West. Asking for the rest-house we were told it was up at Babusar Pass, about 13km (8 miles) further on. Most of this distance was taken up by a seemingly endless series of hairpin bends, up through the conifer forests, but then suddenly we were out on the moorlands. Here the air was crisp and clear, and the sight of large areas of snow was a welcome change after the sweltering heat down at Chilas in the Indus Valley; our 50km (31 mile) drive had brought us up to about 3,700m (12,000ft) from Chilas.

'After a few more hairpin bends we were on the last level to the pass, a low col in the rounded skyline. Silhouetted against the snow were a pair of camels, which I later learned belonged to Afghan refugees going down through the mountains to the plains. Unfortunately we couldn't follow them to the pass, as a couple of days earlier large masses of snow had slipped down the slopes on to the track, making it impassable to jeeps. About a dozen local men were engaged in clearing the snow with shovels, and we all got out of the jeep to talk to them.

'Apparently we had been misdirected; the rest-house was in fact down at Babusar village, and not up here at the pass at all. We got back into the jeep to turn it round, myself in the passenger seat to hold the seismometer, which, despite the fact that it appears to be a heavy metal cylinder, is actually a light relatively delicate instrument. We backed a few metres to a point where a gully ran down the mountainside, making the track slightly wider, and after a seven-point turn we were almost pointing in the right direction. Just one more reverse was needed to allow us to straighten up. However, just at this point the right front wheel suddenly slipped off the edge of the track. The rest of the jeep seemed to have no hesitation in following, and as we toppled forward, the downward slope, stretching away for over 200$m$ (660$ft$), came into view. The jeep seemed to take an age to roll off the track, but there was no way of stopping it. We bounced heavily upside-down, and rather than seeing my entire life flash before me, as is supposed to happen in these situations, I just waited for the bounce which would flatten our roll bars and canvas top, and end it all.

'However, after rolling one and a half times, we suddenly stopped, having fortuitously stuck behind a boulder that itself was lodged in the gully in the mountainside. James and I crawled out through the bent doors. I was practically unscathed, but James had a chipped tooth and a triangular notch in his nose which looked suspiciously similar in shape to one of the three feet on the base of the seismometer, which was eventually found behind one of the seats. All the jeep windows were smashed, and every feature, including doors, bonnet, side panels, seats, was bent out of position.

'After sitting around feeling dazed for a few minutes, we carefully removed all our equipment from the back of the jeep. The car batteries used to power the seismograph were leaking acid everywhere, but miraculously everything else was still in working order. Rehman organised the locals who had been clearing the snow into righting the jeep, but it was clear that we couldn't possibly get it on to the road again. After arranging for porters to take care of the equipment, we walked back down to Babusar village. In order to get James to see David Giles as soon as possible, we hired a local jeep to take us back down to Chilas, while Rehman stayed at Babusar to deal with the jeep. It was eventually lowered down the slope to the next stretch of track, and after some on-the-spot repairs, was carefully driven back to Chilas, and thence to Peshawar for a proper refit.'

It had been a miraculous escape. James had a badly bruised chest, possibly bent ribs, a bloody and dented nose. Despite the fact that he took a calm, detached view of the incident, he was now, understandably, a very cautious man. The vehicle had to be repaired at expedition cost, and Nigel had to do some awkward juggling of the accounts and the schedules of the remaining vehicles in order to keep everyone on the move without loss of effort.

Frankly, I was amazed that the vehicle was repairable at all and even more so when within four weeks we had it back in service. Meanwhile, however, James had yet another lucky escape from an accident involving a vehicle, and his usual cheerful smile and gleeful spirit understandably now waned, never quite to return to their former infectious level. It is to his credit that, in spite of this, he did not once lose his commitment to his work.

Of course, every expedition has its crop of accidents and illnesses which in retrospect are trivial and quickly forgotten, to be recalled only in jest at reunions over pints of once-forbidden drinks. The more serious and tragic accidents always occur on some other expedition. Although we were close to Diran, Rakaposhi, Nanga Parbat and Haramosh, these mountains appeared far too beautiful and quiescent to be the site of major tragedies. It seemed almost unbelievable that Nanga Parbat had killed more climbers than any other major mountain, that Rae Culbert and Bernard Jillott had both died on Haramosh, and that Pakistan commandos had died on Diran while trying to recover the bodies of previously fallen comrades.

Deep down, we knew the risks involved, but having taken all possible precautions to safeguard ourselves, we went about our daily tasks without further thought of a catastrophe. This feature of human behaviour is the strength of mankind, because without it we would achieve nothing and the world would stop. But we were destined not to return without a major tragedy, as I recorded in my diary for mid-July.

'The date is now 20 July and the tragic story of the last five days needs to be recorded.

'I had had a premonition that something was wrong as I walked along the road to meet David Collins coming back from his work on the Batura Glacier. The skies were darkening and very soon the only light would be from the stars, a few of which were now quite bright. I did not want David to be alone working in such a dangerous situation on the banks of the river, and my thoughts concentrated on what would happen in case of an accident at this late time of night. Thankfully I saw David some 2km (1¼ miles) north of the Pasu camp and we walked slowly down the road together, I slightly relieved that nothing untoward had happened. We chatted very amiably all the way back to the hut, where supper was ready to be served. It was a most enjoyable evening, full of banter and good cheer between the members who, having now got to know one another and being more appreciative of each other's roles, were enjoying the intimate atmosphere of the Pasu hut, far away from the more luxurious but tourist atmosphere of the camp at Aliabad.

As we were enjoying our second cup of coffee a Land Rover arrived, the headlights shining through the windows into our dimly-lit single room. Someone said, "Ah well, it's a good excuse to make another cup." Shane and Tom Crompton came into the hut, and Shane immediately sat down next to me, put her arm across my shoulders and said. "I'm terribly sorry, Keith, but there's some sad news to tell you. It's about Jim." My immediate thoughts went out to James Jackson, because of his recent accident. I thought something had happened which had caused him to have a relapse, possibly because of the pressure on his chest when the jeep overturned. I had been quite concerned, as all of us had, over James's accident because he was obviously in some pain and was not willing to admit it. I asked if it was serious. Shane said it appeared to be very, very bad indeed, and very serious. It then dawned upon me that she had said "Jim" and I turned and said, "You don't mean Jim Bishop?" She replied that she did, and that he had fallen from the summit area of Kurkun at 4,730m (15,520ft). The entire room was stunned into silence. No-one spoke. Spoons stopped moving in cups and the primus went out. Shane gave all the details she knew; although death was not confirmed, it was obvious that the accident was very, very serious. Awan [our Pakistan Surveyor] had seen Jim fall down a very steep precipice, for at least 300m (1,000ft) before he disappeared out of sight. The accident had occurred about 5.00 pm on 14 July, Jim's thirtieth birthday. A few seconds before his fall he had told Awan to take great care because the rocks were unstable, a little damp, and therefore slippery. He had been reconnoitring a safe route along the jagged ridge only approximately 150m (500ft) from the very summit of Kurkun itself. Earlier, the swirling mists had caused them to take the wrong route which, although feasible, had created a few extra difficulties; as a result, they had left the bulky survey beacon behind them until they had found a good route to the top. Awan, always just a few metres behind Jim, had heard no call or noise, but instinct made him look ahead; not seeing Jim, he glanced down and saw his body falling down the precipice. No-one could have survived such a fall. Awan had had a difficult time getting down in the darkness after waiting awhile to see if there was any movement from below. He rested a few hours by some shepherds' huts on high pastureland and then in the early morning light he ran down the gorge to the village at its foot to telephone Gilgit and base camp from the construction site of a new irrigation canal.

'After the shattering news sank into my mind, I went for a quiet walk on my own to the end of the white stone road leading from the hut towards the river. I was so shocked I could not help but cry. I had been with Jim on four expeditions to date, and had been in some very tricky situations with him, notably in Greenland during the traverse of the Staunings Alps in 1975, and also during the scientific projects we had conducted on Vatnajökull in Iceland in 1977. The personal loss was so great that it was a little while before I thought about Jim's wife and family; and this upset me further. I then had to sit down and think about the significance of the tragedy to all involved. I immediately recalled all the last-minute hitches we had overcome to enrol Jim into the expedition, and how happy we all were when all the loose threads had been pulled together at the last moment to permit him to rejoin the expedition he had so reluctantly resigned from several months previously. I remembered how Jim had taken the responsibility from me for acquiring the radios only a few days before my departure. These radios were now to prove a great boon in the recovery attempt which Nigel had immediately started organising.

'Jon Walton, Jim's brother-in-law and Antarctic companion, had gone down to Gilgit with Nigel. I now bitterly regretted my selfish impulse to stop at Pasu one extra night rather than journey down with the Pakistani contingent. I quickly packed my bags and Tom and Shane drove me down to Karimabad (see map, p. 100). Tom drove exceedingly careful in the darkness, always in third gear and never faster than 50kph (31mph). At the temporary bridge over the Ghulkin Glacier melt-stream it was almost pitch-black, but fortunately two nightwatchmen were on guard just in case the bridge was loosened. The river was now hitting the ramparts at ninety degrees with full force, and it was obvious that the bridge would not stand much longer. The very steep approach of the vehicle onto the trembling bridge, and the shaky traverse over to the next and last foundation, was very dangerous, especially in the dark. Shane went ahead of the vehicle and directed us over the bridge to lessen any chance of an accident. However, I don't think any of us were too scared; our thoughts were elsewhere.

'The drive down through the gorge towards Karimabad seemed endless. Nobody spoke. What could one say?

'Shortly before the camp site, we stopped the vehicle and I said what I thought we should do. My first desire was to drive down to Gilgit and take charge of the rescue operations at first light, but Shane and Tom thought it would be better if I rested the night at Karimabad and drove down early in the morning. However, on learning that the helicopters were to come into operation at about 4.30 am, I realised that if I delayed now operations would be well under way by the time I got to Gilgit. I was concerned that the wrong information would be transmitted to the UK. It was easy to imagine what trouble and distress could be caused through possibly incorrect information being released by the international press. Previous experience had taught me that it was imperative that all information should be suppressed until we had exact details and a full account of the accident. Obviously, we first had to be one hundred per cent sure that Jim was dead, and this meant locating his body if at all possible. Furthermore, I knew that Ron Ferrari was due to leave from Gilgit for England at any moment, and I wanted to catch him before he left, so that he could take with him an agreed statement for transmission to all parties concerned. He was our best hope of ensuring that the correct information was transmitted home.

'It was some consolation that all the Pakistani members at base had stayed awake until I arrived, and they immediately came over and gave me their condolences. We chatted for a few moments and I thanked them both personally and on behalf of the expedition for their show of kindness. Within a few minutes of arriving at base it was obvious that, irrespective of the state of the road, I should drive to Gilgit. It took a little time to find a jeep and driver, but eventually my baggage, including climbing equipment, was transferred to the jeep, and Riaz, the manager of the Chinar Hotel, and myself sitting alongside the driver started a long journey downhill; it was Riaz who would supervise the accommodation of our rescue groups passing through Gilgit.

'Boulders littered the road, no doubt brought down during the night due to the rapid lowering of the temperature and the recent rain. Such rubble is usually removed by the maintenance crews very early in the morning, and hence daylight travellers are not usually aware of just how much debris collects on the highway during darkness. Once again no one spoke during the journey.

'Eventually we took a short cut through a narrow gouged-out tunnel which led to a rickety bridge across the Hunza river. We then took a narrow track which led to the suspension bridge over the River Gilgit – a short cut to the Chinar Hotel. It was about 2.00 am when we arrived, and everyone was in bed. I was quickly shown to a room where I could get a couple of hours' rest. I could not sleep for thinking about possible rescue alternatives; I should have known that Nigel would have analysed everything and arranged details of operations to perfection.

'Nigel was extremely angry in the morning; apparently the Deputy Commissioner had been totally ineffectual and had requested information from Islamabad before he would take any action. It was only after seeking assistance from the Military that Nigel had eventually been able to approach the helicopter pilots, about 5.00 pm the previous evening, when it was agreed that we should take off at first light in the morning. The pilots were to now give us every possible assistance in the rescue.

'Riaz himself was very upset because he thought that the shepherds could have been mobilised one day earlier to help the rescue effort – they knew the tracks and the shape of the land far better than anyone else. Later we were to learn that they did all they could, indeed they gave us tremendous support during our own operations; in fact, it would not have been possible for them to do more, since Jim was lying on the precipice in a position that only mountaineers could reach. The helicopter pilots were also very upset because apparently a similar incident had occurred with a Japanese climber two or three years previously.

'The first flight on the 16th was taken by David Giles and Jon Walton, followed by Awan and Alan Colvill. They were transported very high up the valley gorge to the highest pastureland where a suitable landing-pad could be found for the helicopter. They then started to climb up a slope immediately behind the pad and after 450*m* (1,500 *ft*) reached the lip of a corrie where they found it was possible to construct another helicopter pad. They immediately radioed this news down to Gilgit, but the helicopters had prearranged business to attend to and were not free until about 4.00 pm, when they took the next load up, including Ashraf Aman, the K2 summitter, and myself. Baggage loads were very critical, since the changing air temperatures and air current flow in the later afternoon meant that only much reduced loads could be carried.

'On the first run the oil filter light blinked warningly just as we neared the lower pad and we had to return to Gilgit rather hurriedly and in an emergency condition, the helicopter skipping over ragged rock needles and swaying a little in the air-blasts coming up the steep walls of the gorge. The pilot had to cut all the corners, and the only fuel tank we could use read empty as we landed in Gilgit, with none of the usual approach formalities. The fault was soon corrected and we took off again, but with a further reduced load. The pilots asked if I wished to see the exact position of the body, since they had now located Jim two-thirds of the way down the precipice. He had fallen into a gully containing two ice patches and was lying face-down, head uppermost near to the foot of the higher patch. It would be no easy matter reaching him, let alone bringing him down.

'I saw no point in going too close to the precipice again. The priority now was to get everyone together as quickly as possible in order to decide immediate actions and to ensure the safety of those attempting the recovery. The helicopter therefore landed at the lower pad first to allow me to gather up further equipment while it took Ashraf on to the upper pad. It then came back for me before taking Awan back down to Gilgit.

'Meeting Ashraf in Gilgit was fortuitous and I was very glad I had requested him to join us. Because of his very high altitude experience, he would need very little acclimatisation and could help in the first bid towards finding a route through towards the body. Awan, meanwhile, had done all that was required and more besides, and it was important that he should now be rested. The Storno radios were excellent and communications between the upper helicopter pad at the corrie edge and Nigel in Gilgit were first-class.

'We talked frankly about what course of action we should follow, bearing in mind the needs of Jim's family, primarily his wife and mother, and also those of the expedition and local authorities. Having discussed all alternatives openly and frankly, I told Jon that I would support him in whatever he thought best. I knew he would never put other people at risk or create a situation worse than our present one. Jon was impressively cool-headed about all these issues and stated that he, too, felt that if any climber had to put his own life in jeopardy in order to recover the body, then regrettably the attempt would have to be abandoned. We both knew that Jim would have wanted it no other way. However, we agreed that, if at all possible, the body should be removed from the mountain and carried down to Gilgit for transmission to Islamabad.

'Throughout the next few days, Dr Giles was an incredible pillar of strength to all, preparing drinks and meals and getting everyone started on time, as well as giving exceedingly wise counsel to all those who asked his opinions. Jon and Ashraf set off very early in the morning and quickly attained the vertical buttress to the south of the corrie lip. The difficulty was to find a high enough traverse line to enter the gully. The lower approach looked far too difficult and would necessitate a steep and dangerous ascent of the lower ice field. They climbed quickly, but had to attempt several alternative routes in order to find the best way through to the upper ice patch.

70

During this ascent four climbers arrived from Gilgit in two helicopter lifts; Ron, Tony, Bob and Brian. All were emotionally shattered. Bob Holmes was not yet fully acclimatised and was unsure of climbing on the very loose rock. Tony, Ron and Brian Whalley needed a rest before attempting the rescue, since they had come up from a very low altitude, approximately 1,200m (4,000ft) to almost 4,300m (14,000ft) in only fifteen minutes. It was necessary for them to acclimatise a little and have some nourishment before starting off on what could be a long, difficult and dangerous mission. Alan Colvill was only just recovering from his early illness, and wisely thought it best not to impede the main thrust but to act as reserve strength.

'By this time Jon and Ashraf had found a satisfactory route through to the gully, to a point slightly above the icefall on which Jim was located. They had fixed a safety line across the most difficult part of a traverse and now was the time that they should return. They had done far more than could be expected under the circumstances.

'I accompanied Tony, Ron and Brian up to the base of the buttress, where they roped up on a single rope of three. I followed more slowly, feeling a little unsure as to what positive role I could now play. Because of lack of sleep and an eye infection, I was probably more of a hindrance than a help. Certainly I had not yet acclimatised, so after a short climb and soon after almost being stoned by a rockfall caused by the rope above, I decided it was more prudent to leave my pack of two small ropes behind in a cache and let those above continue without further interference from me. The climb from here went up shallow grooves beneath a huge shattered boulder, and thence right towards a ridge which took a very easy angle towards a steep buttress. This was traversed by using the fixed rope left by Jon on towards a small snow slab. Once across the slab a short, hard rock pitch led to the ridge that bounded the couloir in which Jim's body lay. Jon and Ashraf were now climbing down, having done all that was requested of them. They were emotionally exhausted. Radio communications on the face were excellent, and hourly schedules were maintained throughout to give moral support to all parties. The second party now recorded that Jim must have had no knowledge whatsoever of the fall and had been killed outright. His body was carefully wrapped in a blanket and it was decided to lower him all the way down the gully that fell steeply away in a series of precipices down the near-vertical north face.

'Every effort was made to recover the body, but unfortunately, when Jim was lowered down one very short precipice where it was hoped that he would land on a small ice field and then gather speed to go over onto the next section, he had other ideas. His body folded over and went straight into the bergschrund (the crevasse between the rock wall and the ice patch) to a depth of about 5m (16ft). This was typical of Jim's behaviour, on two counts. It was as though he were adamant that he wished to be buried in the mountains that he so loved, and also did not wish his friends to be put at further risk. Certainly, it was far too difficult to extricate him from the base of the bergschrund and frankly a more fitting burial place for a mountaineer could not be found.

'The three rescuers now came down the cliff face, very slowly, separately, and well spaced out. Probably they wanted solitude, but all the same I went up to meet each one individually. Even though some had only known Jim for a little while they were heartbroken. We all wept. Brian was too stunned to speak. Ron Charlesworth was deeply upset, and Tony Riley, between tears, recalled witnessing a similar tragic accident involving Don Morrison a few years previously in this area. The doctor, in his deep understanding of human reactions, kept us well fed and coaxed us to have drinks at regular intervals during this very trying period. Alan, Bob, Ali (Sirdar of the porters) and I went back to get the two ropes that had been left in the niche about 30m (100ft) up the face.

'Immediately we got back to camp, it was agreed to build a memorial cairn. Jon made a short inscription on a flat surface of slate and it was decided that I should go to Islamabad to report and have a small metal plaque made that Jon would then fix on the cairn during a later journey to the same peak. There was no need for further helicopters, and we decided to make an early start down the mountain the following morning. A very brief service was held by the cairn. Jon added a few words, thanking every one of the young men. The sorrow in their eyes was so great that I soon dried up. However, we were all convinced that the best solution had been found and that it was far better to leave Jim at rest, peacefully, in the niche behind the ice wall. I went for a quiet walk into the corrie behind the helicopter pad and recalled the happy, glorious days when Jim and I had trudged the Greenland tundra.

'The 3,000$m$ (10,000$ft$) descent was started early the next morning, but we stopped temporarily at the shepherds' pastureland where they had built a few summer huts. We chatted a little with these people, whose feelings were exhibited in their soft, sympathetic eyes. The descent was a typical alpine one, down a series of ribs and rock rognons and steep gullies. Flowers grew in profusion in many places, swayed by the water droplets that escaped from the tumbling waterfalls. On the steeper sections, covered with loose soil of extremely fine texture, we took extra care since the footholds would scarcely support our full weight. Sometimes we traversed high ; sometimes we descended down to river level; sometimes we crossed over from one bank to the other. I frequently stopped to look back at the massive final tower of Kurkun and to drink the cold but clear-tasting water out of the river.

'As we passed shepherds' caves and rock caverns on our right, I couldn't help but recall that Jim had climbed up through this marvellous terrain full of hope and enthusiasm only a few days previously. I felt that it was time to reflect and reminisce, so I urged David Giles who was with me to go ahead more quickly while I journeyed on alone. I reached the bottom of the valley, probably one hour behind everybody else, at about 1.00 pm, the descent having taken about five hours. Nigel Atkinson and Awan came up to meet me, and I took this opportunity of telling Awan that I thought that his actions over the past few days had been very correct, very courageous and very humane. He was obviously a little troubled, and concerned that he had carried out the correct procedures. He really had no need to worry about things, but that was the nature of the man. On reaching the vehicles I was given a bottle of beer, which I shared with my two companions, and then we drove down the track for about an hour until we reached Gilgit.

'The track, which was sometimes hewn out of solid rock, was just wide and high enough for the vehicle to pass along without scraping the rock roof above. At the Chinar Hotel a magnificent meal of steak and chips awaited us all. Nigel Winser had done wonders and was getting transport organised to get everyone back to the comfort of all our friends in Karimabad. We had a short discussion as to who should go back to Islamabad, but I maintained that I should go in case anything went wrong with the arrangements. My previous experience of such accidents made me aware of many possible pitfalls. Under no circumstances should the news be leaked to anyone until Jim's immediate family had been informed of the tragic accident.

'Rana was back in town and reported that the recent rain had created many rockfalls on the Hunza road and that only jeeps could get through to Karimabad. Meanwhile, I talked to Ashraf, who had impressed me greatly during the past three days, and suggested the possibility he join our expedition. Jon, too, thought it a good idea that he should become a member and work in the survey team. Certainly, Jon now needed as much assistance as possible. Some rescue equipment borrowed from the Pakistan Alpine Club at the Nilt

Mountaineering School was repacked, and soon Nigel and myself were alone at the Chinar Hotel. Tickets were booked for me to fly to Islamabad the next day, and we then went round to see James Jackson at the Tourist Inn. He told us of a near-fatal accident that Geoff had had when he found himself driving through the middle of a rockfall. It transpired that he had been extremely lucky that no major boulder had toppled his vehicle over and that only small boulders had landed on the roof as he drove through the avalanche. It was obvious that we were going to be exceedingly lucky if we were able to complete the expedition without any further fatality.

The tragedy of Jim's death remained fresh in our minds for the entire project. It is still unbelievable that one of our most experienced members should have fallen to his death, but the loose nature of the rock, the swirling mists, his concern for the welfare and safety of his companion, no doubt all contributed to the initial slip.

To order a temporary or permanent withdrawal would have been improper, since this would have meant abandoning all that Jim would have wished us to achieve. For Jim and his family's sake, we knew that we should continue and make the expedition a success, if only in scientific terms. In this way, we would secure a worthwhile and everlasting memorial of benefit to the international community of which Jim had not only been a member but to which he had contributed so extensively in his short but active thirty years. On our return home we learned that, before the accident, a glacier in Antarctica had been officially named the 'Bishop Glacier' to commemorate his work in the South Polar Regions.

# From Peak to Peak

In March, 1940, a remarkable meeting took place at the RGS when Eric Shipton reported on his 1939 Karakoram expedition. In the audience were Sir Francis Younghusband, Professor Kenneth Mason and Mr Scott Russell, who was the botanist on the 1939 expedition. Younghusband was seeing for the first time pictures of the area he had explored more than fifty years previously (in 1887 and 1889) and which gave ample evidence to support his claims that the country through which he had travelled resembled 'hundreds of Matterhorns put together,' a statement that had raised the eyebrows of the mountaineering pundits of the European Alps at that time. In his comments, Mason gave credit to the excellent map of the Hispar and Biafo glaciers drawn during the course of the 1939 expedition by Peter Mott, who acted as a consultant to the survey team of the IKP. This map was started from the Hunza Valley using the inter-continental triangulation completed in 1912 and 1913 by Mason. It was this triangulation that also gave the base for what was to be one of the IKP's most exciting investigations (see page 75).

Mason himself had worked south from the Pamirs after linking in with the Asiatic grid that had just been brought south from Osh. The linking of the Asiatic and Indian triangulation network was an achievement of great import, not only because it was the completion of the longest survey over the largest land-mass on earth, but also because it indicated that, once political problems were eliminated, scientists of many nations could combine to achieve worthwhile research. The work was obviously planned and executed at a time when the great powers were in a non-expansive mood and in a sufficiently strong position not to feel threatened. Nevertheless, a cursory glance at page 75 will show that the network goes from Pakistan, into China and then into Russia, avoiding Afghanistan territory.

Mason's report on the Hunza Gorge section, which was the responsibility of Mr V.D.B. Collins, stated: 'The country is extremely difficult. We were observing to light signals, which meant leaving men on the mountain tops, perhaps for a week on end, ready to show a lamp by night or a helio flash on the sun by day. The average height of our stations was (about) 17,000 or 18,000ft [i.e. about 5,300m] but we had to carry up heavy theodolites. When we joined our triangulation with that of the Russians the probable error was only 1.5m (5ft), so that we were not far out.'

What an understatement! To emphasise the difficulties of the area, Mason then went on to tell the story of the Yengutz Har Glacier near Hispar. 'In 1902 it came forward so quickly that, according to native report, it caught up two old ladies who were running in front of it! I do not altogether believe that story, but there is no doubt that it moved forward 2 or 3 miles [3.2 to 4.8km] in about six weeks. It could be seen advancing.' Perhaps the 'two old ladies' were trapped in one of the sudden avalanches of boulders and dust as they scurried over the moraines.

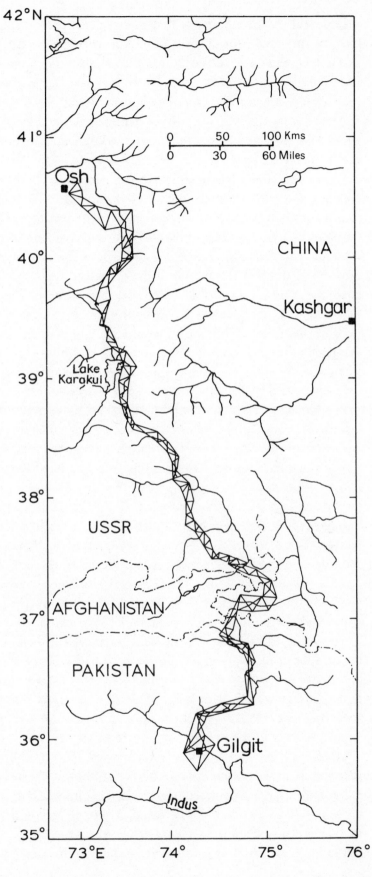

*The triangulation grid established in 1913 between Osh and Gilgit*

Little did those gathered at that meeting realise how much influence they would have on our own Project. Mason had unwittingly crossed over a zone where Asia and India collide, and the accuracy of his painstaking work had provided us with the base data which would enable us to measure the effects of that collision. With modern instruments and new techniques it would be possible not only to complete the same work quickly, but also to achieve even greater accuracy for the benefit of those who may wish to continue this work when a similar time period has elapsed, that is, in AD 2047.

Papers read at the Islamabad conference indicated that the two continents may be colliding at a rate of approximately 15cm (6in) per year, and so it was to be expected that possibly 10m (33ft) of lateral movement would have accumulated in the 67 years since 1913. One aim of the survey group was therefore to determine how much movement had actually occurred, and where. Detailed analyses of possible errors, however small, had shown that if the movement was uniformly distributed, then it could be almost impossible to locate. But since the movement occurs in very localised areas, particularly along fault lines, then there was every possibility of an exciting discovery. This work would be complementary to the seismic studies detailed in the next chapter and was seen as one of the Project's major undertakings. The selection of the Director of the programme was therefore a crucial issue, and no better person than Jon Walton could have been found.

In addition to Ted Smith (who had had valuable experience in the use of lasers in his research work), John Allen (now doing a PhD at Sheffield), and Alan Colvill, Jon enlisted the services of Tom Crompton, a lecturer in survey at University College, London, who despite his lack of mountaineering experience (he had once walked up Snowdon) turned out to be the find of the expedition. Not only was he an expert on his subject and would take on any task and execute it to perfection, but his good humour never deserted him. When I was eavesdropping one day on a radio conversation between the three survey parties on their three different peaks, he interrupted proceedings to report that he had seen my hat walking around base camp, approximately 3,000m (10,000ft) vertically below his theodolite telescope and that 'presumably the boss is back'. I took my cap off and he immediately confirmed the fact. Nothing went undetected. Tom, now Deputy Director of the programme, was a most able theoretician as well as a sound practical surveyor, and he introduced Jon to Nigel Atkinson, who had just completed an MSc in surveying and the year previously had had his trip to the Karakoram cancelled. Jon reported that 'he was exactly the kind of man we wanted; army background, trained surveyor, meticulous, very clued up and "a natural".' Nigel proved to be a born expeditioner who more than lived up to his reputation.

James Jackson had previously mentioned the possibility of a seismic study to Roger Bilham of the Lamont-Doherty Geological Observatory in Palisades, New York, and Roger immediately saw the double importance of a re-survey of Mason's network, which he initially postulated. Although he could only spare a few weeks in the Karakoram, his intimate knowledge of surveying, geology, plate tectonics and seismology made him an ideal choice to help in the location of possible major zones

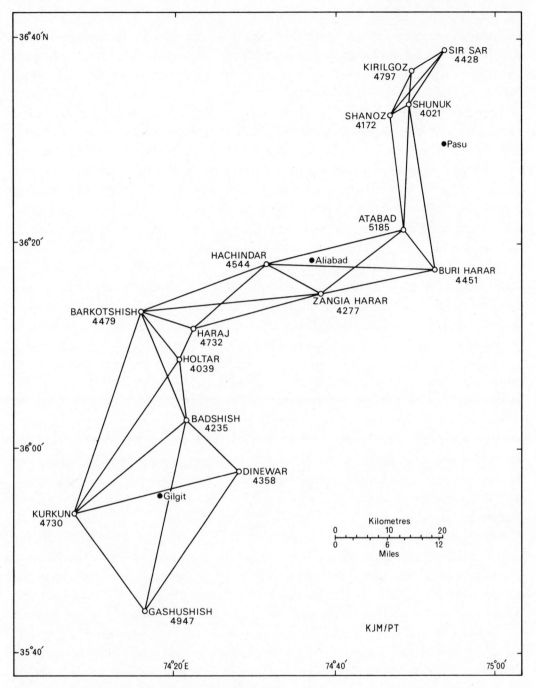

*The section of the grid re-examined by the IKP; heights in metres*

of movement within the network. More will be said about Roger later, but it is interesting to note that he, too, had been dubbed a CIA agent, this time by the Iranians, who thought he had been sent out in 1978 to check on the effect of a mythical Soviet bomb that was supposed to have caused the Tabas earthquake!

A major asset to the team was the inclusion of Chen Jianming from the Lanzhou Institute of Glaciology and Cryopedology, China. Like all the other Chinese, he was very shy and diffident at the start, but he had much recent experience of surveying in the Hunza Gorge, having worked on the Batura Glacier and also on the Karakoram Highway construction project with the Chinese–Pakistani team. Jianming, tall and thin, had an inexhaustible supply of energy. This was just as well, since he set himself an equally inexhaustible number of tasks to perform. With men like him on an expedition, failure seems unthinkable.

Immediately after the tragic accident on Kurkun it was essential that we found another person who would increase the climbing strength of the team -- and who better than Ashraf Aman? He is the only Pakistani to have climbed K2, the second-highest mountain in the world at 8,611*m* (28,250*ft*) and probably the most difficult. Strong and willing, determined and cheerful, he knew the locality exceedingly well as he had been born in Hunza. As well as the local language, he also spoke Urdu and English and was to be of great assistance in a variety of ways. Ashraf had the lovable trait of turning a joke round completely and aiming it at himself, usually amid great laughter. His knowledge of folklore kept everyone entertained for hours. To our amazement, if not consternation, Ashraf would manage to keep up a constant chatter, despite a breathtakingly rapid ascent and the weight of his pack. Indeed, when he introduced himself to Jon, his first words were: 'Hello, my name is Ashraf Aman; my hobbies are walking and talking.' During our attempted recovery of Jim Bishop on Kurkun, Ashraf loaned us much of his own equipment pending the arrival of additional rescue equipment, and on the climb itself exhibited so many endearing qualities that Jon and I blessed our good fortune when Ashraf agreed to join us. The only difficulty was to convince the authorities in Islamabad that Ashraf should be freed from his guiding duties on tourist trips and be permitted to join our project as a full member. However, by this time I did not relish negotiating any further agreements so I quite simply informed the authorities that I had appointed Ashraf as a full member of the expedition. To my relief this *fait accompli* was accepted.

In selecting this colourful but complementary blend of personalities to form his excellent team, Jon had displayed his own natural leadership abilities. And just as Jon had faith in his own team, so we had faith in Jon's ability to organise the programme, although it must be admitted that I was worried that too little time was given to this crucial task. Whereas Andrew Goudie had been planning his programme for almost two years, Jon started in January, only twenty weeks before departure. However, he had by then taken up a part-time appointment at University College, London, and gained tremendous support from his Head of Department, Dr Arthur Allen. Meanwhile, I had authorised John Allen to go on an eight-day course on satellite doppler receivers.

British Industry soon appreciated what Jon Walton was attempting to do; George Wimpey Ltd generously undertook to finance half the costs of this programme, and British Aluminium donated 200*m* (650*ft*) of tubing for the manufacture of 4*m* (13*ft*) high tripod beacons. The British agents for Kern and University College, London both provided geodetic theodolites worth £6,000 each, and the instrument companies of Wild Heerbrugg (UK) Ltd, Tellurometer UK Limited and Aga Geotronics all rallied round with loans of theodolites and electronic distance-measuring equipment, as did Hunting Surveys, a company from which Peter Mott had recently retired from his position as Managing Director. Smaller items such as tripods, heliographs, levels, altimeters and compasses came from the School of Military Survey at Newbury and the RGS Stores; and finally, we were lent, free of charge, the much-valued JMR 1 Satellite Doppler Receiver, which would have cost us £30,000 to buy, or at least £500 per week to rent. After such generosity, we did not begrudge having to find the largest insurance premium ever taken out on survey equipment for an expedition.

Contributions came from varied sources. The team itself contributed almost £3,000, The Royal Institution of Chartered Surveyors another £1,000. Other donations came from the survey firm of Longdin and Browning and also from that Worshipful Company of the City of London which has helped countless expeditions, the Drapers, of whom Kenneth Mason had once been Master. Finally, the Royal Society, who contributed to three of the IKP programmes, gave almost £3,000 to the survey team.

With the modern equipment now at our disposal it would be possible for us to improve on the remarkably accurate survey carried out by Mason in 1912–13; that is, if the stations could be relocated. This problem caused many a grave moment during the planning and execution of the programme, even though Peter Mott told us that two of the stations were in good condition in 1939. Meanwhile, Tom carried out a complex network analysis on a computer to determine the value of the proposed work and to provide an analysis of possible errors and accuracies.

As the excitement mounted, Jim Bishop telephoned me only three weeks before departure to say that his firm, the consultants Sir Alexander Gibb and Partners, were not posting him overseas immediately and was it possible for him to join in. I was absolutely delighted, and soon his firm was as enthusiastic as the team and his family. Jim took over the purchase of the radio equipment, which was to be the last modern lightweight equipment to be added to our stores, and which would make the programme, and indeed the whole Project, the success it was. Radio links were essential, and one of the first jobs Jim took on was to teach us how to use this equipment.

Predictably, virtually no maps were made available to the team, and of the aerial photographs only a few were handed over to us, at a price, and after much delay. However, the Survey of Pakistan loaned us the services of Abdul Razzaq Awan, whose role was to gain as much experience with our modern instruments as possible, a task he performed with surprising ability and obvious delight. Awan was a spritely, energetic man, full of good cheer and completely genuine. Although seldom to hand he always appeared at the crucial moment. His only apparent concern centred on the ability of our batteries to keep the satellite receiver working, perhaps because he was over-keen to demonstrate his electronics wizardry to local villagers. As you can guess, Awan was soon loved by us all and there were poignant scenes when we had to say goodbye at the end of the summer.

From time to time we also had the services of another Survey of Pakistan Officer, Mohamed Farooq, a tall, proud and intelligent man who sometimes worked with

Awan and sometimes with the Housing and Natural Hazards group. He too was keen to extract as much experience as possible from our joint exercises, but was somewhat constrained during the first five weeks of the fieldwork by having to perform other duties designated to him for a summer season in the Northern Areas by his superiors.

Jon led a team that had had many man-years of previous expedition experience in wild and desolate places, where bivouacs rather than base camps provided nightly shelter, and lightweight dehydrated foods, not fresh vegetables and fruit, formed the daily menu. In consequence, his team displayed that indefinable quality of an expedition; a unity of purpose, perhaps more adequately described as an all-for-one and one-for-all philosophy. No complaint was ever voiced by this group, irrespective of the heavy demands placed on them both by themselves and by other project teams.

The survey group was involved in more climbing than any other team (see Figure 7, p. 77). Indeed, during the course of the expedition, it was estimated that the team ascended more than 200,000m (656,000ft) or twenty-two Mount Everests, and contended with a corresponding range of temperatures: 0–35°C (32°–95°F). Because of the steep and unstable slopes that had to be climbed, the members of this programme were highly vulnerable to serious accidents, as the death of Jim Bishop on Kurkun and Alan Colvill's slashed face graphically illustrated. And so when Ted Smith went missing on Hachindar, the concern felt by all members could readily be appreciated.

Ascents to and descents from survey stations were not a simple matter. In this area of 'a hundred Matterhorns', poor visibility could mislead the surveyors and guide them to the wrong line of ascent, perhaps the wrong ridge, and in some circumstances even the wrong peak. To complicate this situation, mountaineers frequently believe they have spotted a better line of ascent or descent than the one previously followed. It was therefore not surprising when Ted Smith, descending from Hachindar, decided to find a better route down into the gorge. When he failed to arrive at the pick-up point, there was no immediate concern; but as the hours passed and darkness fell, doubts increased.

To reassure myself, I remembered Ted's ability to lose himself, as he did once in Greenland in 1973, when the smell of food eventually led him back home – Ted loves his food and hates to miss any meals, scheduled or otherwise. This proved to be the case again this time, in the Karakoram. Of course, he was unmercifully ribbed when he did return, and there were many remarks about short-sightedness, and comments about N on a compass meaning 'North' and not 'Nagar', and how 1700 hours meant a rendezvous at 5.00 pm, not that seventeen hundred hours were allowed for the descent. All this jovial banter, however, hid our relief that no harm had come to him. Ted had, in fact, descended a far more difficult route than he had anticipated.

One of the distinguishing features of the survey work compared to the other projects was that the same work had been done sixty-seven years previously. However, in 1912 and 1913 vast armies of porters had been employed to carry the bulky and heavy equipment and, most important of all, to build platforms for the station on which the equipment would be mounted, and on which tents could be perched for several days until readings had been successfully completed. On some occasions, surveyors had to wait many days for the other parties to reach their positions and transmit the tell-tale helio flash. Jon Walton's team, on the other hand, had of necessity to move fast while maintaining efficiency; they only had the short summer season in which to complete their work. But they did have the great advantage of being able to use radios and direct-measuring instruments to record distances between stations accurately. Mason had to start from a given base line of known length and to determine distances using only observed angles and trigonometrical relationships. In 1980, not only did we have lightweight, robust and exceedingly accurate instruments for measuring the angles between stations, we also had tellurometers that could measure distances to an accuracy of 1cm (0.4in) per 1km (0.6 mile), and a laser that could measure to the same accuracy over 10km (6 miles). This meant that we were able immediately to check on all distances calculated from the initial base line and also to construct a series of base lines of great accuracy up the gorge itself. The use of radios also permitted communication between survey parties; if line-of-sight communication was not possible during ascents to peaks or on the summits themselves, then an intermediate transmission could be facilitated via base camp at Aliabad in Hunza.

To illustrate the advantage of this, one story is well worth telling. Jon had already located Shunuk station, but Tom was reporting no luck in finding Shanoz (see Figure 7, p. 77). It could well have been located on any one of a number of spires on the ridge ahead, and so it was by no means surprising that he had chosen the wrong peak. At the appointed time of the radio schedule, Jon looked through his own theodolite telescope and reported that this was not the station required but one some hours' walk further on along the ridge – that is, unless the continents were colliding at a rate of knots! Since these peaks are neither easily nor quickly climbed, Tom's laconic reply was: 'Same radio schedule tomorrow – over and out.' This degree of synchronisation was the hallmark of the magnificent teamwork of this group, although in the first three weeks they had several moments of doubt.

The first task was to survey the most northerly series of triangles from Pasu. The Liaison Officer, Major Rana, went with the group to ensure that no one strayed beyond the officially set limits, although I did not interpret these limits too literally since Roger Bilham had located a possible major fault in this neighbourhood that could have accumulated some deformation. Placing science first and politics second, I told Major Rana of this decision, but instructed him to report fully on what we achieved because by this time it was clear that he had to make regular reports on our activities to his superiors. This first survey was entirely successful and the four teams returned to base in four days, elated by the ease with which they had accomplished their tasks.

They had wisely chosen to work on a quadrilateral of small sides. From Pasu they progressed to the village of Khaibar and from here up to the stations of Shanoz, Shunuk, Kirilgoz and Sir Sar. The local commanding officer of the FWO assisted the team, particularly with the selection of porters, and from this experience they set their future pattern of selecting porters from the villages close to the stations. As a result they became pleasurably acquainted with local lifestyles, customs and cuisine.

Four stations had now been located at the corners of the quadrilateral, beacons erected and sightings and distances recorded, and it was necessary to plan the next stage of the survey. The group now discussed where and when to place a further batch of beacons in relation to the stations from which the surveyors would operate. From these decisions it was necessary to work out the composition and equipment of the various teams and when they should start, so that each team would arrive at its station simultaneously. Finally, pre-arranged radio schedules had to be carefully worked out. Obviously, with no intimate advance knowledge of the routes, no-one could judge if an ascent would take two or three days and if ten or twenty porters would be required to carry food and equipment to high camps. We also had to allow for camps to be stocked for several days, since it was impossible to guarantee that the 4m (13ft) high beacons would be clearly seen in conditions of variable visibility through theodolite telescopes over distances of 30 to 40km (18 to 24 miles). On one occasion Jon, accompanied by the film crew of Tony Riley, Ron Charlesworth and Bob Holmes, ascended to within 100m (330ft) of the summit of Haraj to erect a beacon, but with daylight running out rapidly, they had been forced to return to the valley 3,000m (10,000ft) below without achieving anything. It was the same day that they learned that Jim Bishop, on a similar mission, had fallen on Kurkun. Jon was heartbroken: Jim was his brother-in-law; they had worked together in Antarctica, and the love and respect they had for each other was plain for all to see. At that moment the expedition became a family, and we all had to watch and assist one another through the difficult days ahead. Jon, however, had the worst job of all. He had to remain steadfast and loyal, courageous and sensible while still directing his group's activities; only in a few lonely moments could he permit his private emotions to be released. Jon eventually returned to Kurkun and erected the beacon that Jim had left only a few metres from the summit. Later on, he returned again to take the necessary survey readings.

The southerly end of the network did not go as easily as the northerly start. Jon had a 32km (20 mile) walk followed by a stiff climb up to Gashushish station at 4,947m (16,230ft), and he and his team of Jianming and porters, including Ali, the Sirdar (chief) of the porters, arrived at the top some two and a half days out from Gilgit with only a few minutes to spare before the pre-arranged 10 am radio call.

They only had food for one more day, and the next day a short period of bad weather was sufficient to interrupt their work drastically. Eventually, although they managed to measure one triangulation line, put up one beacon and measure one vertical angle, they had no choice but to return disconsolate to Gilgit.

There, further bad news awaited. Alan Colvill had been taken up the wrong ridge by his porters and was suffering severe dehydration. Meanwhile, Tom had ascended Dinewar and had sat around waiting for the others to appear on some of the myriad spires at various points of the compass; when at last they did appear, the battery of his tellurometer had run dry. The same problem beset Nigel Atkinson when he climbed Chamuri. His journey, with only one porter, was some 80*km* (50 miles) from Gilgit by road and donkey, and then they had to climb 3,000*m* (10,000*ft*), up an unknown route. After such a magnificent effort he, too, found his tellurometer had been damaged when the donkey carrying it had slipped and fallen. In spite of this major setback, Jon rallied his forces and again they set out, Jon and Ted Smith up the ill-fated Kurkun, Tom and Jianming up Dinewar, Alan up Badshish – the correct summit this time – and Nigel to his peak far to the south.

The story of Tom Crompton's return to Dinewar is worth recording. He had left all his equipment, including tent, sleeping bag and food, safely cached in a shepherd's hut, with the shepherd's guarantee that it would be safely guarded. Unfortunately, Tom could not return on the appointed day, and so by sign language he had despatched his porters with the information that he 'would not ascend today, but would return tomorrow instead'. Unfortunately, in translation this became: 'No, he would not ascend, instead tomorrow he would retreat.' Accordingly, when he returned to the village, Tom was met by a posse who informed him that his porters had just ascended to the high camp in order to bring down all his equipment. The porters had no sooner entered the village – in a state of near-collapse, since they were then into Ramadan, the month of fasting – than Tom had to persuade them to return in haste back up the mountain in order to keep the radio schedules. Tom's sense of responsibility was fortuitously assisted by his sense of fun and also by his close friendship with the local population, which saved the day. The hard trek back was started with a minimum of delay, a few laughs and a face-saving bonus payment.

Thus, in seven days this group ascended and descended 18,300*m* (60,000*ft*), this incredible effort being rewarded with a satisfactory series of readings and, most important, proof that our beacons could be sighted over a distance of 35*km* (22 miles) without too much difficulty.

From then on, Jon reports, 'it was plain sailing; hard work but plain sailing. Of course, there were difficulties, as when I collapsed from dehydration and had to be carried up the last 150*ft* into camp' – this during the ascent of Badshish. As was to be expected, there were occasional altercations with porters over pay, but this is an occupational hazard of all expeditions. In general the team's memorable stories reflect the friendship, love and humour that were the rewards of working with the indigenous population. Never to be forgotten was the porter who walked a great distance down the valley stream every day to catch trout for the evening meal; and the dog that stole the chapattis one night. This latter incident took place during Ramadan, the month of fasting for all devout Muslims, when it is essential that food be eaten only during the hours before dawn and well after sunset. Of course, on the tops of mountains the sun appears early and only disappears long after the valleys and gorges are sheathed in dark shadows. Thus the porters, having worked hard and long during the ascent, had to bake their bread well into the night, ready for eating at about 3.30 am the next day – stray dogs permitting.

As work progressed, the team became more and more impressed by the quality of the stations built by the 1913 surveyors, but an apparent discontinuity occurred on

Hachindar, which had been used in the intermediate years by surveyors from the Survey of Pakistan. It was quickly realised that the wrong station had been used, since it was not intervisible from the other key points. The 1913 station was quickly located some 2km (1.25 miles) further back along the ridge. This solved the riddle of computing errors in the Survey of Pakistan's analysis, which can now be corrected since the 'wrong' and the 'right' stations have been precisely co-ordinated into the network.

The major epic of the survey work, however, was reserved for those peaks closest to and in view of our base camp in Hunza at Aliabad. From here, parties had ascended Hachindar, north of the river, and Zangia Harar, south of the river. A third party including Jon, Jianming and Ashraf were bound for the most inspiring of all the stations, Atabad at 5,185m (17,011ft). This station was placed on top of an 18m (60ft) pinnacle, and great patience, nerve and delicate footwork were required when moving around and across the precariously positioned theodolite tripod legs. For four days the weather was bad, and time passed slowly. Frustration mounted. In these conditions it was unwise to approach the pinnacle – assuming of course that it could be found.

Meanwhile, the group camped at a picturesque shepherds' gathering ground called Baldihell at an altitude of about 4,000m (13,000ft), where cows, sheep and goats are brought up for summer grazing. The larger animals stay for only a matter of days, the smaller for several weeks before being driven back down the steep alpine tracks. The shepherd told the party that if they sacrificed a goat, then good weather would return immediately. Naturally he had a goat for sale, and so the purchase was quickly arranged and the sacrifice made. The feast was prepared with a minimum of delay. No doubt those who enjoyed their supper that evening gave ample thought to their colleagues, for whom they were doing this great favour. All the remaining groups on the surrounding peaks had to eat that night were near-unpalatable meals made from reconstituted powders of various uninspiring colours.

The next day it seemed the gods were satisfied: the weather was perfect, and the pinnacle was found – much to everyone's delight, since it had eluded detection by surveyors of the Survey of Pakistan for many years. The station was in good order although the route to it was somewhat tricky. In all, sixteen stations had now been visited and in the course of this work, the survey had crossed at least two fault lines.

Another important survey project that was being conducted simultaneously with the re-triangulation was the positioning of selected stations by the team, using the expensive satellite equipment and receiver. John Allen and Awan were initially responsible for the precise positioning of the seismic stations. Given enough satellite orbits it was possible to do this to an accuracy of better than 5m (16ft) in longitude, latitude and altitude. Later Awan was joined by his colleague Mohamed Farooq, and they became so proficient that they were permitted by Jon to go off into the wild and operate the equipment on their own. At the termination of the programme their record books showed just how well-organised they had been, and of course, they had been delighted by the responsibility given them and had responded accordingly.

As September approached, the survey team split up. They deserved a holiday, and

The mess tent at Aliabad
base camp.

Zangia Harar, one of the
survey stations.

Chen Jianming with an
electronic distance-
measuring instrument
on Atabad.·

Ted Smith eating again!

several members took a two-week trip up the Batura Glacier. Jon and Tom, however, responded stoically to my request to help out on the Ghulkin Glacier survey, and they and I had our own busman's holiday, a tale to be told later in this story. Not content with the sustained effort required for their own studies and the additional work associated with the glaciological research, they also worked alongside the earthquake group *and* managed to squeeze into the last few days a traversing programme for the geomorphologists related to the levels of old lake-beds on the Hunza river above base camp.

Overall, the survey team displayed the most unselfish and prolonged dedication to work that I have ever witnessed – an achievement to complement the efforts of Nigel and Shane at base camp, who kept all groups of the IKP well supplied with food, transport and technical supplies. The final task was for Ashraf, Jon and several porters to carry up 100*kg* (220*lb*) of cement and 250*kg* (550*lb*) of sand to the corrie just below the final pyramid of Kurkun and the point where Jim lay at rest. Here, with the aid of local shepherds, a magnificent cairn was built as a memorial to our friend who had given all that he could to the success of the Karakoram Project.

# Earthquakes

In 1951 a group of visitors to the fort at Baltit, the 600-year-old building in Hunza from which past Mirs had planned plundering raids and exercised autocratic and absolute power, heard a deep, subterranean rumble which caused them to run out onto the castle roof. This proud building, the most historic in the Karakoram, was being subjected to a substantial earthquake. Little comfort could be found on the roof, however, since the tall building is perched on a high knoll and surrounded by high precipices.

There followed three hard shocks a minute apart, with continuous shock waves between. The old castle swayed and creaked, but it seemed to be so flexible as to be almost earthquake-proof. As the earthquake stopped, a new and spectacular scene developed as tremendous avalanches thundered down every gorge in the vicinity, and for a further five minutes a continuous rumble was heard and felt. A great snow cloud spilled out and over the sides and mouth of the gully to the east. This particular avalanche was timed at 100*kph* (62*mph*), a wild, tumbling mass of cold air in which powdered snow was suspended. When the 600*m* (2,000*ft*) high avalanche cloud struck, the castle shivered again and everyone had to lie flat to avoid being blown off the roof, while the avalanche continued to roll down out of the dark grey storm clouds above.

The duration, strength and aftermath of such earthquakes bring home the fact that man is not the master of his environment and that he is but a transient on the surface of a world in constant upheaval. Even a small earthquake is an unforgettable experience. An initial sense of wonder turns to helplessness and panic. You can no longer trust your eyesight as your brain cannot register a true horizontal; as a result it is impossible to stand upright. A primitive response is to act like a drunkard and fall to the ground in order to make a more secure contact with what should be solid earth. But the earth vibrates up and down and sways from side to side. Even birds fly in fear, suddenly deprived of refuge. A feeling of physical sickness is not uncommon in large earthquakes, since co-ordination between the senses of sight and touch is disrupted, and new smells and fearful sounds add to the growing sense of inevitable destruction. All these sensations are magnified inside villages, towns and cities when buildings collapse and fires start. Often there is no time to crawl away from potential death traps and it is not unusual for entire populations to be trapped and die from fear, thirst, suffocation, fire, or simply by being crushed.

In the villages of the Karakoram, those working in the fields during an earthquake watch the surrounding landscape for avalanching cliffs and boulders. Those indoors seek out the strong wooden pillars or doorways for protection.

Fortunately, the Karakoram is one of the least populated areas on earth and large earthquakes cause few deaths compared to those in other parts of the world. But one result is that this zone is given little attention, although some of the world's severest earthquakes originate here and are coupled with some of the most diverse land forms on earth. The reason for the large release of energy here in the Karakoram, as demonstrated by the frequency and intensity of earthquakes, is that for the past forty million years the Indian subcontinent has been drifting northward and pushing against the Eurasian land-mass. Before the collision took place, the Tibetan plateau had a maximum altitude of only 1,000m (3,300ft) and a climate to match. Hippo, giraffe and buffalo freely roamed the country. Today the average height is 5,000m (16,400ft) and the plateau is almost totally devoid of any form of life.

The theory of continental drift has its origins in the eighteenth century. Theodar Lilienthal, a German theologian, pointed out in 1756 that the facing coasts of many continents (e.g. western Africa and eastern South America) would fit back together, but the theory of continental drift was first popularised by Alfred Wegener in 1910. Wegener visualised the continents as thick plates, floating on top of the thin and weak material that supports and contains the oceans. However, it now appears that continental drift originates in movements in the sea floor, caused by the slow but continuous welling-up of material from deep within the earth along fairly well-defined mid-oceanic zones. The modern theory of plate tectonics is thus a much broader concept than the original continental drift theory and provides a fairly simple explanation of the large-scale motion of separate parts of the earth's crust. It postulates that a small number of large plates belonging to the high-strength outer shell of the earth move rigidly with respect to one another, at rates of 1 to 20cm ($\frac{1}{2}$ to 8in) per year, over the hotter and lower-strength material beneath. The outer material, the lithosphere, is about 100km (62 miles) thick and consists of the earth's crust and the upper part of the mantle. In consequence, the crust is being carried along as part of the plate (see page 88). It follows that boundaries of tectonic plates can be under the ocean or across land-masses. Thus, this theory simultaneously considers both the land-masses and the other part of the earth's crust which lies beneath the oceans. The oceanic crust is relatively thin, about 6km (4 miles), whilst the continental crust is approximately 35km (22 miles) thick. The oceanic crust is constantly being replaced by new material coming up from the mantle in the rift of the mid-oceanic ridges, and this keeps a balance with that part of the oceanic crust that is being drawn down, or sub-ducted, into the asthenosphere. The asthenosphere is hotter, less viscous and weaker than the lithospheric plates, and convection currents within it are an important driving-force for plate motion. Continental crust itself is both thicker and more buoyant than the oceanic crust and so cannot be drawn down into the dense mantle.

Where one plate passes under another, mountains can form at the continental plate boundaries; the Andes of South America, as well as the Himalaya and the Karakoram are examples of this process. The theory of plate tectonics also indicates that the oceanic crust is replaced every 200 million years, whilst the thicker continents have existed for billions of years and have been deformed many times.

The detailed movement of the Indian continent shown on page 89 has been plotted

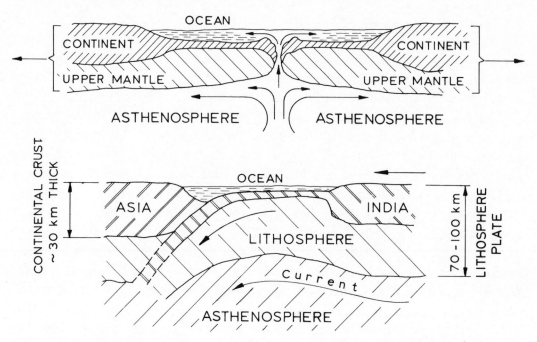

*Convection currents deep within the earth are the principle driving force in continental drift.* ABOVE *The oceanic crust is constantly being replaced by new material at the mid-oceanic ridges.* BELOW *In the collision of India with Asia, the Indian plate is being forced under the Asiatic plate*

from a knowledge of when and where the crust was formed. Age can be determined by various techniques, including isotope analysis, while position can be deduced from the magnetic character of the new crust as it solidified. In consequence, the Indian plate is known to have moved 5,000*km* (3,100 miles) in a north-easterly direction for 30 million years before full impact with Asia, and it has moved another 2,000*km* (1,250 miles) during the past 40 million years.

Some evidence of this is provided by the fact that few Asian fossils approximately 45 million years old have been found in India, and that the fossils a little older than this show strong affinities with those of a land-mass known as Gondwanaland. Fossils younger than 40 million years include those of plants and mammals that migrated from the Asian plate when the two continents were joined.

However, there is also other evidence that indicates that India was quite close to Eurasia 500 to 600 million years ago. Thus the geologically recent collision between the Indian continent and Asia has given rise to complex questions about its underlying causes as well as resulting in the formation of the great mountain ranges previously described. The zone of primary interest for us was where the two continents collided and sutured, or stuck, themselves together. Suture zones can be difficult to locate, especially in the Karakoram, but Tahirkheli and Qasim Jan had previously helped define two possible collision zones in an area of some geological complexity.

This complicated area was one of great significance to the IKP, since it incorporated an arc of island volcanoes (referred to as an 'island arc') similar to those of Japan or the Aleutian Islands. This line of volcanoes lay close to the Asian

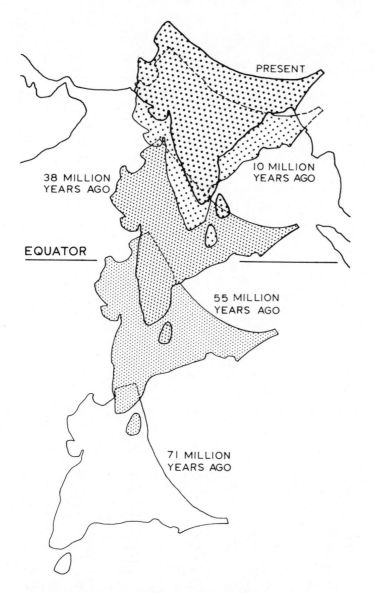

PRESENT

10 MILLION
YEARS AGO

38 MILLION
YEARS AGO

EQUATOR

55 MILLION
YEARS AGO

71 MILLION
YEARS AGO

*The collision path of the Indian continent with Asia*

land-mass, rising out of the oceanic crust, and was crushed by the Indian plate during the collision with Asia and slowly raised and tilted along the axis shown on page 90. In consequence, it is now possible to travel from the surface of the earth down towards the centre of the earth simply by travelling south from Hunza through Gilgit to Patan, along the Karakoram Highway. This is one of the most exciting of recent geological discoveries. There, some 2,000km (1,250 miles) from the sea, not only can one find evidence of this uptilted island arc and information relating to the forces of creation, but – more spectacular from the mountaineer's viewpoint – it is possible to stand on the road near Gilgit and see Rakaposhi to the north, 7,788m (25,550ft), possibly the highest point of an island volcano, close to the Eurasian plate, whilst to the south, moving towards it, is Nanga Parbat, 8,126m (26,660ft), standing on the Indian continent peninsula like the bows of an advancing ship.

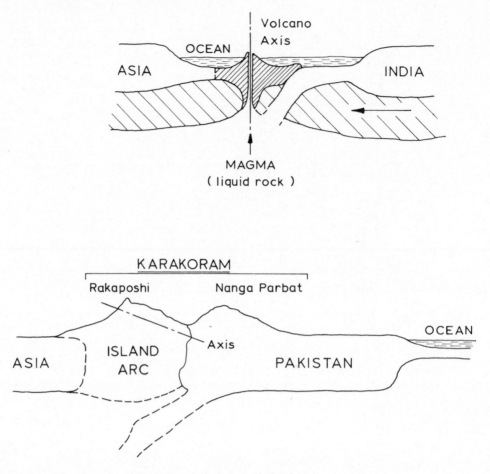

*The tilting of the island arc in the collision between India and Asia*

Perhaps, after all, it is not so surprising that the theories developed by mechanical engineers to account for the deformation of hot metal being squeezed in rolling mills can also be applied to the deformation behaviour of these large continental masses. However, calculations of the rate at which the colliding continents and the trapped island arc are being squeezed and the highest mountains on earth formed show that deformation has accounted for only a small fraction of the movement since the collision. The riddle to be solved, therefore, is: how did the Indian continent continue to move at a general rate of about 5*cm* (about 2*in*) per year, which apparently it has been doing for the past 40 million years, giving a current displacement of 2,000*km* (1,250 miles), and how was this massive movement accommodated within the continental mass?

An answer to this intriguing question may be found from satellite photographs, which show several types of land-mass movement along faults during earthquakes. Indeed, the clearest of all faults are to be found here in Asia. An analysis of the direction and extent of movement of the different blocks of the continent, particularly at the Altyn Tagh Fault, north of the Karakoram, which continues into another great fault to the east known as the Kansu Fault, shows that China is being pushed eastwards. The combined fault is probably the longest and greatest horizontal slip system in the world.

The study of earthquakes is of great importance, since they not only help to indicate the exact localities of deformation, but they also give information on the type of geological structures through which the tremors pass. Furthermore, earthquakes are evidence of large instantaneous movements between different geological structures, and the damage they create is immense. From information recorded on seismographs, it is possible to model the structure within the lithospheric plates and to determine the planes of weakness and possible zones for future earthquakes, thus hopefully pinpointing future areas of potential catastrophe. As long ago as 1556 a single earthquake near Sian, the then capital of China, is thought to have killed 830,000 people; in the summer of 1976 the great earthquake that devastated the Chinese industrial city of Tang Shan was the result of forces originating in the Indian/Eurasian plate collision area, some 2,500$km$ (1,550 miles) to the south-west. This earthquake is reputed to have killed another 400,000 people.

Early discussions in Cambridge with Dr James Jackson revealed to me the valuable contribution that could be made to the Karakoram Project by the inclusion of a study on earthquakes. A knowledge of the precise position and strength of tremors would provide data on faults and geological structures, information that could be used to model collision zones and to indicate their points of weakness. Such a study would also complement the surveying, housing and natural hazards, and geomorphology programmes. At the outset it was hoped that the seismologists would operate north and south of the Karakoram, that is, in both Chinese and Pakistan territory, via the Khunjerab Pass on the Karakoram Highway.

Soon the team grew to involve friends of James who also worked in the Geodesy and Geophysics Department at Cambridge University, namely, Dr Geoff King and a research student, Graham Yielding. The British membership was completed by the inclusion of Dr Ian Davison of Leeds University. James and Geoff became Joint Directors of the programme, so that decisions could always be taken during the planning stages when one or the other would be away in some corner of the world, examining the aftermath of an earthquake. They both had much experience in operating under difficult conditions and in consequence thought that they could cope with all the various types of dangers man and nature could contrive. As will be seen, however, all the team witnessed events well beyond their previous experience. This had the effect of diminishing Geoff's naturally aggressive nature to one less brusque and more compliant to the environment in which he had to direct his team's effort, and this was to help him successfully through to the termination of a deeply committed and sustained effort of the most arduous kind imaginable. James, on the other hand, again because of the considerable danger attached to the mission coupled with its prolonged nature, lost his usual exuberant, happy and carefree nature to become quiet and, with justification, nervous, if not a little frightened. Nevertheless he, too, like all other members of this team, never lost his determination to collect all the data that time would permit.

The international membership of the team was almost completed when three

Pakistanis, Shabbir Ahmad, Akbar Khurshid and Rehman Khattack were selected. Shabbir is a lecturer in the Earth Sciences Department of Quaid-i-Azam University, Akbar is a geophysicist working for the Geological Survey of Pakistan, and Rehman is an electrical engineer working with the Pakistan Atomic Energy Commission.

As a result of my negotiations in Beijing, Lin Ban Zuo joined us from the Institute of Geophysics of the Chinese Academy of Sciences to complete this team, and he too proved to be a most valuable member of the Project.

To everyone's surprise, no member of this unit was permanently maimed during the fieldwork, but there were several narrow escapes. It was reasonable to expect a high accident rate since each and every member of the team was constantly forced to drive long distances over the worst terrain for jeeps imaginable. Furthermore, throughout the entire period of the Project, the group had little if any relaxation, primarily due to the fact that they only had one task: to record, on drums blackened by carbon, the tremors of earthquakes reaching the surface of the earth. Unlike the other teams, they had no other study to offer them a respite.

At the end of the expedition, all members looked like scarecrows. Between them, using Land Rovers and jeeps, they had travelled several thousand kilometres across high deserts and steep precipices, under loose cliffs and glacier snouts and through rivers and avalanching boulders on tracks likely to subside, if not totally collapse, at any instant.

Many of the earthquakes recorded were of the order of tens of kilometres deep, and of these, most were of low intensity and not felt by humans. Only modern, highly sensitive transducers can sense these deep tremors and transfer them onto the delicate recording drums. The records of these shocks, together with records of the same earth tremor recorded on other seismographs at far removed sites, enable the position of the earthquake to be pinpointed in a manner similar to survey triangulation techniques. In order to achieve great accuracy in the locating of an earthquake, it is necessary to have an array of instruments spread over a wide area, each picking up the tremor. Obviously, the more seismographs recording an individual tremor, the greater the accuracy in pinpointing the source. From data such as this it is possible to model the collision zones and make predictions of future trends. This information, together with the work of the surveyors and geologists, helps us to establish whether the mountains of the Karakoram are still being displaced horizontally and vertically, as well as telling us where to expect future earthquake activity.

One of the team's major problems was to find good sites for the recorders. They could not be situated close to rivers or villages, since even the soft footsteps of the investigators were picked up by the transducers. In order to avoid this, and to eliminate vibrations from passing vehicles and rolling stones, sites had to be found which were at once remote from extraneous noise, of geological interest, readily accessible so as to permit frequent maintenance, and above all, safe. In the Karakoram there are few such localities, and the team spent a not inconsiderable time examining possible and impossible sites. In general it was easier to find sites higher up the valleys, where the rivers had less flow. The final sites of the stations are shown below.

The group organised their own schedules, visiting outlying stations from Chilas and the Tourist Cottage of Gilgit, the latter under the management of Mr Abdul Karim. This able and very intelligent entrepreneur was trained to operate the local Gilgit station while the teams were away servicing other stations. Karim also provided a room for the seismic group with a variable number of beds which were used by all our itinerant wanderers. His excellently prepared food was served under the stars, in the relaxed atmosphere of a soirée. Life at the Tourist Cottage was not hard, and it was used as a resting-place following accidents, bouts of Chilasi fever and general fatigue.

During the Eid festival to celebrate the end of Ramadan, which was attended for several days by many Mohammedan members of our project, James made a trip to Astor, close to Nanga Parbat (see Figure, p. 52), which appeared to be an active seismic area. However, siting a seismic station there proved very difficult because the Astor river (the old route to Gilgit from Srinagar) is like a slot through the mountains and it was impossible to find a quiet location away from the river, down which boulders were constantly trundling. Consequently, the seismic data was poor.

The lower Astor river is awesome as it flows through dry and barren débris, but further upstream and approaching the south-east side of Nanga Parbat, everything changes: here, it rains. At Rama, near Astor, there are lush, green

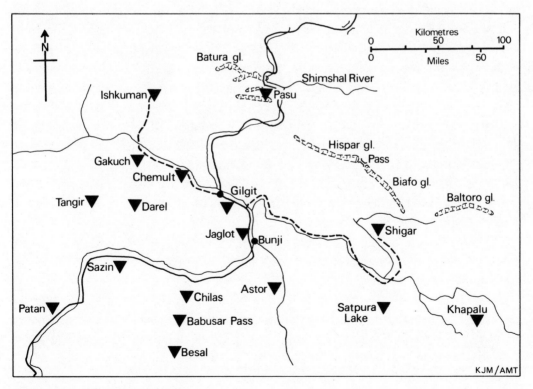

*The location of the seismic stations*

meadows and pine forests beneath the snow fields and ice walls of Nanga Parbat, and instead of mica-laden glacier melt-water, there are clear streams full of trout. In this quiet location we placed a seismic station.

James and Akbar Khurshid later went to Baltistan to run three instruments, with the intention of monitoring valleys which looked like active faults on the satellite photographs. A branch of the Karakoram Highway now goes to Skardu and is an easy, though spectacular drive, following a narrow thunderous gorge of the Indus, more impressive than that near Patan. However, the road to Khapalu beyond Skardu is hideous beyond belief.

James arrived at one desolate section early one morning to find that the road was a mere 60cm (2ft) wide shelf in the cliff, 30m (100ft) above a very angry-looking Indus river. A kilometre back down the road, a construction crew, when woken from their slumbers, said it would be repaired in three hours – a statement that would have strained the credulity of even the most flamboyant optimist, since the road simply did not exist for 100m (328ft) and was merely a miserable ledge covered with enormous boulders. It was barely possible to walk along it, let alone drive. But after breakfast, as if from nowhere, about fifty people appeared – jeep drivers who had been waiting patiently on the other side of the cliff section and had now climbed round to offer their services. Using nothing more elaborate than crowbars, these people rebuilt the road in three and a half hours by constructing a dry stone wall on a tiny projection 6m (20ft) below and building it up to the level of the road. According to James, it was a privilege to witness this feat, which was performed with great cheerfulness and applause as they heaved great boulders into the Indus. The prima donnas of the operation were the people who balanced the stones on the ledge below the road, perched like sparrows, without ropes. Had they fallen, their fate would have been the same as the boulders! The final product was horrific; a wobbly, bumpy surface which just did not look wide enough for a jeep; James was assured that it was safe since it had been built by the same people who intended to drive over it, but after previous experiences this argument seemed as fragile as the road itself. The first jeeps went over, with much shouting and waving of directions. When the time came for James to cross, he adopted the tactics of the UNMOGIP (United Nations Military Observer Group in India and Pakistan) and walked over whilst his vehicle was very slowly driven across, to an accompaniment of anxious cries and shouts. The driver grinned and recommended that the return journey be done at night, since the darkness would conceal the risks.

To minimise the possibility of an accident, Akbar remained at Khapalu whilst James operated two stations out of Skardu, one up the Shigar Valley to the north, one south at Satpura Lake.

Satpura Lake is idyllically quiet and blue. The only person living nearby was the chokidar (watchman) of a rest house, Jaffir, who spoke English, had a great sense of humour and became James's companion and guide. The lake yielded up trout every day for a month. However, food in Skardu was a problem, since the cook's repertoire was limited to incandescent vegetable curries, and no eggs could be purchased. Consequently, James settled into a routine of eating trout at Satpura and eggs at Shigar, a village en route to K2 which provides a delightful mixture of sand dunes and peach orchards.

Of all the seismograph sites, Tangir was probably one of the better ones but that was all that could be said about it. Let Graham Yielding tell the story.

'Geoff and I had installed one of our instruments in the northern part of the Tangir Valley, on a seemingly remote, lightly wooded hillside. When we returned four days later to change the recording paper, the instrument had disappeared; the recorder, the geophone, the batteries and every single piece of connecting cable – everything had gone. Theft of the batteries, which were identical to those used in jeeps, was understandable, but to remove all the equipment seemed too much like hard work for the average thief, especially as it was doubtful that anyone within miles would have any electronic knowledge pertaining to the instrument. We drove back to Sutil, the main village in the Tangir Valley, and went to the house of the Malik (the local head man). He and his son had been very helpful in establishing our station in Tangir, and he agreed that this was a very serious crime, and that we ought to report it to the Chief Inspector of Police. First, however, he insisted that we had lunch. The Chief Inspector was very interested in the crime. "This is the first time we have ever had a theft reported in Tangir Valley. Normally we just have murders – fifteen per year – but never have we had a theft."

'The scene of the crime had to be visited, and Geoff offered to drive the Inspector there in our Land Rover. Almost as an afterthought, the Inspector decided we ought to have some protection, and a number of policemen, armed with fast-repeating rifles, were bundled into the back of our vehicle. At the village nearest to our former seismic station, the Inspector gathered together the elders and a meeting was convened in front of the mosque. For much of the afternoon they argued back and forth in the local dialect. Eventually, it was decided that the instrument would be put back anonymously, and would await our return to Tangir. A few days later, upon the next visit, the instrument was still missing. The Malik told us that this was probably because we were his friends and that he was not especially liked in this particular village. A reward was then offered, and after another few days the instrument was produced, although without the jeep batteries.

'Our general feeling about these villagers was such that we felt it would make future visits to the seismic station easier if we repositioned it further up the valley, where the people in fact turned out to be quite friendly. We subsequently discovered that when the instrument was first put back it contained a letter, written in Urdu, which said that our work of "earthquake prediction" was an interference with the will of Allah and hence a blasphemy. Any attempt by us to continue using the instrument in the same place would result in our death. However, the police had removed this letter before letting us have the instrument back, so as not to offend us. The fortuitous repositioning of the seismic station further up the valley was therefore in the best interests of our health.'

Establishing each station brought its own individual problems. Ian Davison informed me that when he and Geoff first installed the seismic station at Gakuch, they assumed that the local authorities would already have been informed about our activities by the Commissioner in Gilgit, so they quietly set up their instrument without advertising the fact, reasoning that the fewer the people aware of the instrument, the better. They left a sign on the instrument which implied that anyone who touched it would receive 10,000 volts in return. This sign was removed soon after their departure. When Ian returned four days later to change the recording drum, he was accosted on his way to the station by the chokidar of the local rest house, who asked if he was a Russian. Apparently, soon after they had set up the instrument, a suspicious villager had informed the local police that there was an electric device hidden in the rocks near their village, and two police officers had quickly set off to investigate. Since they had been unable to ascertain what the instrument was, an army expert had been called in to examine it. Apart from establishing that it was not a radio, the technician had been baffled by it. Two armed officers were then stationed in a concealed spot near the instrument, in the hope that the alien spy who had left it there would return during the hours of darkness and fall unwittingly into the hands of the law.

The next day, word had spread to the effect that a Russian spying device was in the vicinity. Fortunately, some of the locals who had watched Geoff and Ian install the instrument came to hear of this rumour and informed the police that their round-the-clock vigil was unnecessary and that the instrument did not belong to the Russians. What might have happened had Ian paid a visit to the site during the hours of darkness to change the seismic record – a common enough occurrence – hardly bears thinking about. Thereafter, the installation of all other instruments was very well publicised!

Suspicions ran high, especially in those areas close to Chilas where the villagers are renowned for their religious fervour. One team was stopped by people who once again accused them of being 'bad men' because they were interfering with the will of Allah and insisted the work should stop. Ian, however, replied: 'If you are accusing us of being evil because we are trying to predict earthquakes, is not every single person guilty of similar offences when they look up at the sky and try to predict when it will rain?'

His reply had a placating effect, especially since no superior argument could be advanced. Nevertheless, such incidents highlighted the power of local mullahs, who could quickly incite their followers to the point of expelling us. The locals frequently take the law into their own hands and never report such minor issues to the police. However, they do inform the police of any murders, of which there were fifty-six in 1979 – this in a valley containing 40,000 people.

The villages in Darel and Tangir (see page 93) are usually small, as each family has its own. Every village is complete with a watchtower where all the family can take refuge when another village family lays siege. In fact, in these two valleys no house standing alone is more than 46m (150ft) from a watchtower. The menfolk always carry around old Enfield rifles used in World War I; those who cannot afford a rifle have a scythe or an axe, and the poorest people carry a heavy wooden club. Only the mullahs and the children walk around unarmed. The people seem to treat life as of minor importance in comparison to honour or religion, and the method of burial reflects their low regard for human life: graves consist of a shallow depression covered with boards, which are soon uncovered by the wild dogs, who leave the graveyards looking like fields of bleached bones.

Geoff organised several people to help him maintain many 'safe' stations, and

Shane gave him invaluable service, enabling him to concentrate his efforts on stations that were more remote, where constant liaison was needed with the highly suspicious and often threatening villagers. To do this, he had to operate out of Chilas, which since the late nineteeth century has had a reputation as the most notorious village of the entire area.

Perhaps it was the climate that made the environment seem so hostile. Temperatures frequently exceeded 40°C (104°F) at midday, and the evening temperatures seldom fell below 18°C (64°F). When it rained, the humidity levels rose dramatically, and in Geoff's words, 'the mountains started to fall down'. Indeed, on one occasion, there were a series of mudflows, one of which buried the seismometer stationed at Chilas for ever. Whatever the reasons, the Chilasis have always been thought of by most European, Japanese and American expeditions as a mean and vindictive people who simply cannot be trusted under any circumstances. They had a habit of trying to sell the most basic foods at prices that were exorbitant even by the standards of other remote villages, and often much of what they sold turned out to be inedible due to poor storage. Spices were rare, but a little sugar could sometimes be bought on the black market. It is hardly surprising that the posting of a government official to this area, however low the rank, is considered by him to be a personal catastrophe.

Eventually, Geoff's physical and mental exertions took their toll and he succumbed to a high fever, probably malaria, and was forced to take an unaccustomed rest at Gilgit. At this juncture Shabbir from Quaid-i-Azam University took over the reins. He did a magnificent job, and not only kept all the stations running efficiently, but also raised the morale of the team after so many early disasters.

Lin Ban Zuo remained with Shabbir in Chilas to brave the tiger mosquitoes and the heat. Lin proved to be a paragon of patience and an extremely fine cook, even with the most meagre of vegetables, and played a large part in maintaining morale with his chicken casseroles *à la chinoise*. He was also a very able English student and made a great effort to communicate his vast knowledge of seismology to us all. Although a man of mild exterior, he was as tough as leather (as were all our Chinese colleagues), but his toughness was tempered on the one hand by a very funny sense of humour and on the other by a graceful athleticism, as witnessed by his simple but energetic early-morning exercises. On one occasion Ban Zuo, on listening to a punk rock tape, consulted his dictionary and amused us all greatly by asking whether it was western opera.

Whilst Shabbir and Ban Zuo were in Chilas, Ian took an instrument up to the Besal Valley in the Kaghan Valley for a two-week break from the claustrophobic atmosphere of Chilas. He too had been under a tremendous strain, as he was one of only two members of the local team (James being in Skardu) who could drive. This idyllic valley with its temperate climate is the summer home of the Chilasi bourgeoisie, but now accommodates many Afghan refugees. Besal used to be on the 'main' road from Rawalpindi to Gilgit, but is now by-passed by the Karakoram Highway. It remains one of the most unspoilt and beautiful valleys in north-west Pakistan.

All members of the seismic team now knew that their objectives would be attained, and in retrospect it seemed fortunate that it had not been possible to have stations inside China, since the logistical problems would have been far too great. Furthermore, complementary geological studies involving the island arc were now totally absorbing the team, particularly Ian, Tahirkheli and Ghazanfar Abbas, a senior geologist, who joined us for a few weeks from the Geological Survey of Pakistan.

When Geoff recovered, he visited the Ishkuman Valley, since he was particularly anxious to get a station operating as near to the Afghanistan border as possible, and so link in with the 200*km* (125 miles) deep earthquake systems (*i.e.* inside the mantle) known to be operative in the Hindu Kush. This location would also cover the area north of the Ghizar river, close to where the housing group had witnessed a mud flash, and at the same time provide better angular coverage on the deep earthquakes that were suspected to be coming from the northern parts of the Karakoram.

In contrast to the other places he had visited, Geoff found the people of Ishkuman to be kind and friendly, not unlike the Hunzakuts. Their food was superior and more varied, including butter, honey and fresh bread to supplement the peaches, apricots and apples. The only blemishes in an otherwise idyllic scene were the bed bugs, and the plight of the proud Afghan refugees, who nonetheless received their gifts from the Gilgit Commissioner, Wazin Zuda Abdul Qayyum Khan, with great dignity. The District Commissioner himself was widely respected by the villagers for his great knowledge and his considerable ability in local languages.

Although this general locality was comparatively safe, there were two final incidents that reminded Geoff never to relax his alertness. First, the track to one of the instruments was washed away, and he had to wait until it was repaired; second, he came across a lone Japanese traveller who foolishly had visited a high glacier without a local guide. Geoff found him in a state of collapse beside the road, suffering from malnutrition and a severe cold. The lesson is that no-one, however strong, should venture alone into these wild areas at any time of the year.

# The Destruction of the Earth's Surface

Landforms in the Karakoram are developing at a rate faster than anywhere else on earth. Rivers are frequently blocked by mud flashes or rockfalls sweeping down the hillsides and across the valley floors, or by advancing glaciers. Such events cause extensive lakes to form, which can stretch back up the valley from the blockage for many kilometres. Many of these dams are surprisingly strong, since the mud of which they are formed contains stones and boulders ranging in size from pebbles to blocks several metres in diameter; these help bind the whole together just as cement and grit combine to form concrete.

The continual disintegration of the Karakoram mountains occurs because of the interaction of a number of different factors, in particular the effects of climatic variations and the forces of gravity. Rain, snow, ice, frost etc., all play key roles in the general destruction of this mountain area, and when parts of the valley walls break away, or when the rivers undercut these steepest precipices on earth, then gravity causes small and large fragments, in single pieces or in thunderous avalanches, to descend at frightening speeds to the floor of the valley.

The study of landforms in the Karakoram is complicated by the fact that the area is not geologically stable. As has already been explained, this range stands close to the boundaries of two tectonic plates and the pressures created are still pushing the mountains up at the same time as erosion by glaciers, rivers and rain etc. is wearing them down. The geological instability of the area is clearly shown by the numerous earthquakes and hot springs recorded in the Karakoram. The geomorphology team was therefore faced with a unique opportunity to examine the rate and processes of land destruction and also to interact with the seismic group, who were accumulating evidence on the underlying tectonic forces. Broadly speaking, the seismic team studied the causes of the present configuration of the land's surface, while the geomorphological team examined the processes of denudation and the effects of time on patterns of relief.

The team that Andrew Goudie assembled was a good one. Denys Brunsden, a cheerful enthusiast from King's College, London, and David Jones, his hard-working companion from the London School of Economics, had worked together in many countries, including Nepal, and were very experienced at geomorphological mapping. Alayne Perrott had worked alongside me in Greenland in 1972, leading a women's expedition to the Schuchert Valley, and had also worked extensively both in the American Rockies and at high altitude in Ethiopia, making a special study of glacial fluctuations and climatic changes. Brian Whalley had been to the Hindu Kush and was particularly interested in studying high-altitude rock-weathering processes. He was a good climber, and we had been friends during his student days at Cambridge. Broadly the same could be said of Rob Ferguson. He had been to

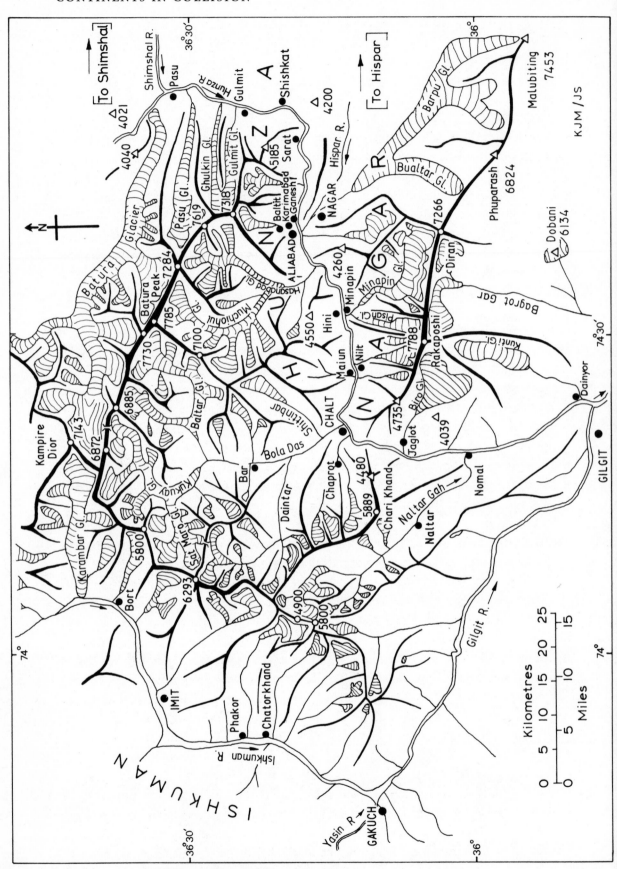

A sketch map of Gilgit, Hunza and Nagar

Professor Tahirkheli, leader of the Pakistani team.

Hunza before, knew a lot about river channel patterns and had considerable general alpine experience. David Collins had worked on glaciers in many parts of the world, notably in Canada and Switzerland, and was an easy-going, cheerful, industrious worker. Another important geomorphologist was Ron Waters, Professor of Geography at Sheffield, who was very knowledgeable about periglaciology, that is, the study of areas close to and affected by glaciers. He had had experience in New Zealand and also had the advantage of having an ex-research student of his, Mohammed Said, as a Professor of Geomorphology at the University of Peshawar. The last UK member of the team, but certainly not the least, was Ed Derbyshire. He too had very widespread experience in the study of sediments and glacial deposits and had been working in China at Lanzhou for five months before his arrival in Hunza, which had given him experience of a similar environment to the Karakoram. He had also worked extensively in Antarctica.

Unquestionably we had a good team of people who would make a determined effort on the various programmes, which had been put together so as to form a complementary series of studies. A pleasing and distinguishing characteristic of this group was that the team came from a wide range of universities: Andrew and Alayne came from Oxford, Denys and David (Jones) came from two different colleges of the University of London, Brian came from Queen's Belfast, Rob from Stirling, Ed came from the University of Keele, while David (Collins) came from Manchester and Ron from Sheffield.

The group, the first to be assembled, had several pre-expedition get-togethers to discuss tactics and strategies. During this period their wide-ranging abilities and experiences were welded together to form a team that was both self-stimulating and capable of working harmoniously with the other developing programmes.

The geomorphologists had a large number of projects to attempt. This had the advantage that alternatives would be available should one series of experiments prove to be impossible, but the disadvantage that it would not be possible to focus undivided attention on one, single-purpose project, as was the case with some groups. Fortunately, all programmes had been carefully selected and all proved highly successful, perhaps because of the extensive pre-expedition planning period. It is expected that some twenty-five to thirty scientific papers will stem from the efforts of this team once the results of the project are finally compiled. This vast quantity of work was in part due to the substantial efforts of our Chinese colleagues, about whom more will be said later, and also to two Pakistani scientists, Professor Said from the Department of Geography of Peshawar University and Zafar Hashmat, a relatively junior geologist from the Water and Power Development Authority in Lahore.

The first project, to survey the snout positions of as many Hunza glaciers as possible, was planned by Andrew. There have been several surveys made of the glacier snouts since the 1880s, the most notable being by Kenneth Mason, the first Professor of Geography at Oxford, who in 1913, as a Royal Engineer, had surveyed some of the glaciers which entered the Hunza Valley. Andrew, being an Oxford geographer, felt it was his 'right and duty, as it were, in memory of the old chap, to re-survey his work; that is what he would have intended.'

The purpose of such periodic surveys is to be able to pronounce on glacier fluctuations, to determine periods of advance and retreat and to assess quantitatively just how much material the glacier has pushed and deposited into the valley. Glaciers are the greatest forces on earth when it comes to transporting bulk material, and it is vital to be able to plot their movements when analysing the decay of mountain ranges.

The Hewlett Packard Company loaned the team an instrument called a 'distomat' which automatically measures distance; a digital flash-up on a screen records how far away a particular target is situated. The target is made of three prisms held on a tripod that is easily transported to key locations. The team also had theodolites and plane tables for mapping. Thus armed, Andrew, Denys and David Jones visited the snouts of the Minapin, the Pisan, the Gulmit, the Hasanabad and Ghulkin glaciers (see p. 100) and set about surveying them. Because of the clear air, the snouts often looked deceptively close, and for the first day or two they found walking up to the glaciers carrying their relatively heavy equipment unexpectedly tiring. They, like every other member of the expedition, had to pass through the acclimatisation barrier. In later surveys they made greater use of porters, which reduced this problem and also left more time for scientific studies.

One of the unanswered questions concerning the glaciers of the Karakoram is just how frequently, and to what degree, they change in response to climatic variations, one indication being the advance or retreat of their snouts. The Minapin Glacier has been well surveyed in the past. It was visited in 1889, when a man called Ahmed Ali Khan made a reconnaissance map, and since then approximately fifteen studies have been made of its snout position. As a result, it is known that between 1889 and 1913 it advanced by approximately 1,700m (5,570ft) – a substantial movement for a glacier. Since then it appears to have been in a general state of retreat; it had retreated by approximately 60m (200ft) since it was surveyed by a Cambridge expedition in 1961, and also thinned quite a lot. The Pisan Glacier has not been so extensively surveyed, but appears to have retreated by 300 to 400m (1,000 to 1,300ft) since air photos were taken of it by the Pakistan Air Force in 1966.

The Gulmit Glacier and the Pisan come down towards the Hunza river, the latter from Rakaposhi. The location of a series of very fresh moraines on the Gulmit gave evidence that at some point in the recent past this glacier had advanced beyond its present position by as much as 1,000m (3,300ft) – probably within the last century or so. Although the team did not have a great deal of material to analyse, they did produce a good map of the snout of this glacier which will be of use to future workers.

The spectacular advances and retreats of the Hasanabad Glacier, which is on the northern side of the Hunza river, make it one of the most extraordinary glaciers on the face of the earth. At the turn of the century it seems to have advanced by about 9km (5.6 miles) in one winter and spring season, but by 1929 it had retreated by 400m (1,300ft), by 1935 by about 600m (2,000ft), and by 1954 by no less than 7km (4.5 miles) from its position at the turn of the century. Indeed, by 1954 the

Hasanabad had retreated so far that it had split into two major branches up different valleys. By the time we went back in 1980, these two branches had come together once again and their new joint snout appeared to be advancing, although it has not yet advanced as far as it did at the turn of the century. Since 1954, it appears to have advanced between 1.5 and 4.8*km* (1 and 3 miles).

The other glacier examined in detail was the Ghulkin Glacier. Records of its movements are limited, but it is of great importance since it is close to the Karakoram Highway and changes in the position and shape of the snout cause a change in the direction of the melt-water stream issuing from it. This in turn can cause serious disruption to traffic along the nearby road. It was a nerve-racking experience driving through the Ghulkin Glacier melt-water stream, as all traces of the Karakoram Highway had been removed. Every so often there was a sickening lurch and bump as the vehicle hit, or was hit by, large boulders that were being driven down by the stream, and Andrew recorded how the water sometimes came right over the bonnet of the Land Rover and seeped through the side doors. Some indication of the real value of the Land Rovers was given by the fact that although we crossed the melt-water stream many times per week, not a single vehicle or person suffered damage or injury. This was not the case with the Pakistani jeeps, which had their sumps ripped out by boulders in the bed of the stream.

Many journeys had to be made across the stream just where it joins the Hunza river in order to carry out the mapping project on the Ghulkin Glacier. In 1913, Mason found that this glacier snout was about 300*m* (1,000*ft*) from the bank of the Hunza river, and we found it in a similar position; but probably in 1966 or thereabouts the glacier was considerably further back, perhaps of the order of 500*m* (1,600*ft*) – and in 1925 probably as much as 1,000*m* (3,300*ft*). So the Ghulkin does seem to have advanced after the 1920s and there is some evidence to suggest that it might continue advancing, which will cause immense problems to the highway engineers.

The temporary blockage and removal of road sections is of major concern, but glacier advances which block off the whole valley are of even greater importance. The river is then dammed, and the resulting lake drowns any low-level road for many kilometres upstream. By the time the dam collapses, the road could well be buried under several metres of new lake-bed sediment.

The second project of the geomorphological team concerned the study of longer-term fluctuations in the glaciers of the Hunza Valley and its tributaries, and this meant mapping the current and ancient moraines of these glaciers from air photos. As part of this programme, Alayne Perrott, assisted by Zafar Hashmat, studied the nature of the boulders on the surface of the moraine, since as the boulders get older, so their characteristics change. Such changes include the degree of surface weathering, the degree of desert varnish development, the amount of lichen cover and changes in the general hardness of the rock. Alayne also studied to what extent soil had developed in the moraines. She and several colleagues spent a lot of time measuring these changes, which enabled them to classify the moraines into various age groups. While this technique permits relative dating, it does not give an absolute age, so we were always on the look-out for old wood, logs and the like in old moraines that we would be able to use for radio-carbon dating. Very little information of this sort has come out of the Karakoram, or indeed the Himalaya as a whole, and so we were very fortunate in being able to find and retrieve several wood samples. These rare finds created great excitement among the geomorphologists.

The third geomorphological project was to survey and core a lake which we had located from the air photographs between the Pasu and Ghulkin glaciers. This lake, called Borit Gil (or Borit Jheel) was first surveyed using an Avon inflatable boat, and its contours determined from soundings; much of this work was done by Chen Jianming, our Chinese surveyor. It was obviously a good lake from which to extract a core, and this project led to one of the more exciting but difficult experiments with which we were concerned. We wanted this core because, in a closed lake basin such as this, the continuous sedimentation processes and the nature of the sedimentation vary according to parameters such as the local vegetation and the presence of nearby glaciers. In other words, the analysis of the sediment in a core taken from the lake floor can give an indication of environmental changes in the area. We hoped that if we could retrieve a core it would demonstrate the continuous changes in organic content and salinity of the lake water and in the pollen of the vegetation around the lake, thus revealing climatic changes. As far as we are aware, no decently dated core has ever been obtained in the huge area between the mountains of Iran in the west and Irian Jaya, Indonesia, in the east.

Borit Gil was a most suitable lake in this respect because of its unique position, and we therefore set about erecting a stable platform from which to undertake the coring. Fortunately, there were three old army pontoons already on the lake, which we named *Hunza*, *Hunt* and *Hemming*, and these formed the base of our coring platform. In order to be able to extract a core, we also needed a long length of pipe which could be let down from the pontoon and into which we could pass our coring machine. Now, in Britain one would just go to a shop and buy some plastic drainpiping; but Borit Gil is in a desert area and here people do not bother with drainpipes – certainly not lightweight plastic ones which are easy to handle. For once, on asking for some irrigation water pipes, we obtained a rapid and positive response from the government officials in Gilgit, who supplied us with rather heavy steel pipes of a large diameter. Unfortunately, these turned out to be so heavy that when they were loaded into the Land Rovers, the springs almost touched the floor. We were worried about taking these vehicles, with their long, protruding loads, up the very narrow and steep tracks to the lake, since the 'road' was really a track for jeeps, not Land Rovers, and it was necessary to reverse countless times round several bends, on surfaces seemingly composed of frictionless pebbles that gave a very inadequate grip.

Eventually the various lengths of pipe were carried from the track down to the lake, where we had to fix four lengths together to make one pipe 12*m* (40*ft*) long in order to reach the bottom of the lake. Mounting such a large and heavy pipe and lowering it vertically into the water was a major problem that was successfully overcome using large amounts of rope and judiciously deploying the pontoon boats.

The pontoons were then fixed into position in more or less the deepest part of the lake by four anchors made from large boulders and tablets of slate through which holes had been punched to attach ropes. The four heavy anchors kept the pontoons very firmly located in the middle of the lake, and taking the core then became a matter of muscle power. Denys was helped by a couple of the locals, who were very keen to join in this unusual European sport of taking muck from the most inaccessible location. Finally we managed to get a 2.28m (7.5ft) sediment core out of the bottom of the lake and this is now being analysed.

This project in itself was exciting, but our enjoyment was enhanced by the fact that the surroundings could not have been better. The lake attracted numerous wading birds, but although the water was clear and inviting, it was not drinkable, largely because of its high content of epsom salts. The lake and its environs were hemmed in by large moraines with a delightful village by one side that housed some of the nicest people we met during our visit to Hunza. It appeared that the further you went up the Hunza Valley, the more friendly the people became. All the local children here joined in with helping us in the task of carrying equipment up and down, and also brought us apricots and apples. As well as being invariably helpful and courteous, like children everywhere they were delightfully inquisitive.

A fourth project involved the study of variations in the sedimentological and geotechnical properties of the various deposits laid down in the area by glacier streams, mudflows and gravity falls. This work was the responsibility of Ron Waters, Ed Derbyshire, Alayne Perrott and two Chinese scientists, Li Jijun and Xu Shuying, from the Institute of Glaciology and Cryopedology at Lanzhou, where all disciplines related to cold regions are studied. Ed, Jijun and Shuying had been working together for the previous five months in various parts of China, including the Lushan mountains south of the Yangtze and the Tian Shan in far western Sinkiang. They knew each other well by this time, and obviously had a good working relationship, which was reinforced by the easy-going nature of Ron and Alayne. Their particular interest was in glacial sediments and in the use of information provided by the sediments to reconstruct former ice extents and determine patterns of glacial behaviour. They had come fresh from doing similar work in the Tian Shan, and within twenty-four hours of reaching Hunza base camp at Aliabad they were on their way to Pasu. The idea was to review quickly the sediment work done on the Batura Glacier by other Chinese scientists in the 1970s, and then to extend the area of mapping and analysis as far down the Hunza valley as time would permit.

Unfortunately, the first fieldwork spell was cut very short by Ed's collapse with a high fever, caused by an intestinal parasite picked up in China. The melt-water river of the Ghulkin Glacier had cut away the road by this time and so the invalid had to be carried across the formidable melt-water stream in the company of the expedition doctor, David Giles, who had crossed the glacier in the late evening half-light to examine the now very sick Ed. Five days later the three sedimentologists (including Ed) were back at work along with Ron, Alayne and Mohammed Said, whose interests were more strictly geomorphological.

The Batura Glacier area (see map on page 100) is representative of much of the upper Hunza Valley in presenting a landscape made up of complex interrelated sediments; originating from a variety of erosional processes and transported to their present positions by ice, water, wind and gravity. One of the tasks was to discriminate between stony silts laid down by the glacier (deposits known as 'till') and stony silts deposited, for example, by mudflows and intermittent streams resulting from the seasonal heavy snowmelt and infrequent but sometimes torrential rainfall in this substantially naked desert landscape. The two types of stony silt are intimately related, but proper classification was essential if the history of geologically recent changes in the area was to be determined. There had already been some academic differences of opinion on this issue before the arrival of the Chinese group, and the question was repeatedly posed during the mapping programme. The mudflow deposits at the base of ravines and close to the valley floor formed the majority of the huge sloping fans, which were inclined at about 6 to 10 degrees and sometimes measured several kilometres across. Locally, the thick fans were interbedded with the deposits from streams which had from time to time cut into them. In some instances, these huge débris-flow and stream fans half-engulfed old moraine ridges left behind by the Batura Glacier. Quite a mélange.

Studies of the moraines of the Batura Glacier led the team to the view that the basic classification published by the Chinese in 1976 was sound. This dated moraines in a relative way from the present back to the Little Ice Age (sixteenth to nineteenth centuries), Neoglacial (about 3,000 to 5,000 years ago) and the last Pleistocene glaciation (over 10,000 years ago). Alayne and Ron spent much time measuring the surface boulders of the till and the desert varnish on the boulders, so as to provide a quantitative test of the Chinese chronology. In sedimentological terms, the bulk of the tills in question were of the type that were let down from the débris-covered glacier surface as the ice slowly melted. However, there were several instances of shearing of till by over-riding ice, and there was abundant evidence that the Batura Glacier ice front had oscillated in the late glacial, Neoglacial, Little Ice Age, and in very recent time.

The 'Pasu' group of geomorphologists were particularly interested in the abundant evidence of temporary lakes in the upper Hunza region. As recently as the 1970s, the Hunza river had been dammed by a catastrophic slope collapse that had led to the formation of a series of transient lakes. There was also abundant evidence that all areas of the valley had contained lakes in the past. Sometimes, as on the north side of the Batura Glacier, it appeared that lakes were impounded by the glaciers themselves as they advanced across the valley and dammed the river. This phenomenon certainly occurred in the case of the Pasu Glacier at the end of the last glaciation, and again as late as the Little Ice Age. Simultaneous advances of the Batura and Pasu glaciers have had the effect of impounding water in the long valley reach between them. We know this happened in late glacial time from the differences in the shape and composition of the melt-water gravels. Thus, the melt-water deposits provide important clues to the history of the area, and at least three outwash plain levels can be traced back into the Batura Glacier moraines. Each level differs in the average particle size, fabric and particle shape of the deposits, with the highest level made up of coarse, rather angular stones, while the lower levels consist of finer, more rounded stones.

There was also evidence to show that, since the end of the Ice Age, huge débris fans several cubic kilometres in volume had grown and engulfed the glacier deposits. The sedimentary evidence makes it clear, however, that in this tectonically-active area débris-fan accumulation and glacial deposition were contemporaneous, the glaciers pushing forward at one time to inhibit the growth of the fans and incorporate their débris, while at others the glaciers shrank and the fans enlarged themselves, reaching the rivers and the ice fronts in broad, low-angled sweeping surfaces.

The geomorphologists and sedimentologists pursued the glacier deposits uphill, and great thicknesses of till were found over 1,000m (3,300ft) above the valley floor, all now eroded into great earth pillars and covered with desert varnish. Opposite and above the Pasu Glacier, these landmarks can be seen at a height of 4,000m (13,000ft). Indeed, from the remnants on the mountains between the Pasu and Ghulkin glaciers, Jijun and Shuying showed that the ice must have filled these great valleys during the glaciations of the Pleistocene. They do not look like typical glacial valleys now, because they have been subjected to such vigorous erosion since; but the record of glacial till remnants is unequivocal. Above the highest till, scattered glacial boulders ('erratics') were found resting on surfaces still bearing the streamlined form inherited from glacial erosion. The ice at such very high levels may have been a part of an ice cap over sections of the Karakoram divide. As they gathered all this evidence, our two Chinese geomorphologists seemed to be as at home on the high slopes as Marco Polo sheep. The Ghulkin Glacier also received close scrutiny from the geomorphologists, but this story will be told later.

The sedimentologists established that the processes and products of glacial action and glacial melt-waters have been broadly similar over the past 20,000 years or so, although the glaciers have been larger in the past than they are at present. They also extended the relative chronology of glacial deposition established by the Chinese back in time, by recognising a tillite, a cemented till with the strength of sandstone, which may date from an older Pleistocene glaciation.

A surface sediment map was completed by the sedimentologists, showing the distribution of the main types of sediment over an area from the Batura ice-front to a point downstream of the Gulmit Glacier. A great amount of sediment was sent back to the UK for laboratory analysis and for comparison with similar glacial deposits from other parts of the world.

A fifth project was to monitor the quantity of sediment and dissolved salts that were coming down the Hunza river and its main tributaries. The reason for this work is that each year the Indus carries approximately 500,000,000 tons of solid material down towards the sea, much of which is today collecting behind dams such as Tarbela, the greatest earthen dam in the world. A substantial proportion of the sediment is coming from the Hunza river, and we wanted to determine where the silt originated. This required very extensive sampling, with the samples having to be carefully treated and then sent to the UK for analysis.

Our studies showed that the land surface in the upper Hunza Valley is being lowered at a rate of somewhere between 5,000 and 10,000 tons per square kilometre per year, a rate of denudation which may be as high as 3*m* (10*ft*) every 1,000 years – extremely high indeed, by any world standards. For comparison, the Thames Valley rate is approximately 5*cm* (2*in*) per 1,000 years. David Collins' work on this project caused much amusement to the locals, who could not understand why we worked twenty-four-hour shifts in order to place very heavy and awkward equipment into very dangerous rivers, simply to take out a small sample that was not drunk, used to irrigate, panned for gold, but just put into bottles and labelled. David did not help matters by wearing what some would described as a perpetual, idiotic grin as he went about his bizarre work. Brian and Rob became deeply involved in this project, which was also assisted by Zafar Hashmat.

Another and novel way of determining the rate of erosion in the area was to measure the extent, distribution and thickness of lake deposits which had accumulated behind the great natural landslide dam at Sarat in the Hunza Valley in 1857. This dam had lasted for a known period of time and, so we argued, by determining the size of the sediment wedge which had accumulated in the natural lake behind the dam, it should be possible to calculate the amount of sediment that was being carried by the Hunza river during that time period. In this project, the geomorphologists were greatly assisted by the surveyors directed by Jon Walton. The results of this programme are awaited with great interest.

It is not difficult to speculate why the Hunza river carries such a phenomenal amount of material. Andrew postulates that it is simply because here is the greatest available relief on the face of the earth. Looking at the summit of Rakaposhi, at 7,788*m* (25,550*ft*), and then casting one's eyes down to the Hunza river itself, the difference in altitude is somewhere in excess of 5,800*m* (19,000*ft*); but the horizontal distance over which this change in altitude occurs is only between 10 and 12*km* (6 and 8 miles). There is nowhere else in the world where this magnitude of available relief exists. Furthermore, it is repeated all the way up the Hunza Valley and its tributaries. This in itself must have a great influence on the rate of denudation. Also, the area is composed of greatly contorted, fractured and broken-up rock resulting from the intense tectonic action in the area and from the action of widespread salt weathering, and of frost weathering at high altitude. In consequence, there is always abundant débris available for the frequent mudflows and landslides. We ourselves often witnessed rockfalls and landslides, and these of course are bringing material down into the rivers all the time, but particularly after rainfall. In the summer months, the discharge of a river like the Hunza increases phenomenally – often by 50 to 100 times the flow of the winter months – as the glaciers melt in the sun. In winter the Hunza river can be easily forded, whereas in the summer months it is a gigantic river of turgid, micaceous mud. Finally, the area has accumulated great masses of material which have been dumped in the higher valleys by its glaciers, particularly during the Ice Age when the glaciers were more extensive. Since this material is friable and easily eroded, it forms an important component of the débris carried by the Hunza river.

A connected project undertaken by the geomorphological team involved studying the mass-movement of material in landslides and mudflows. These phenomena were very frequent, and several members of the expedition witnessed major occurrences. Driving along the Karakoram Highway one day, we saw a great cloud of dust and people running towards us. Andrew, who was driving, reacted very rapidly, slamming on the brakes of the Land Rover. There, in front of the nose of the vehicle, appeared a large section of cliff that had just collapsed. On another occasion, a group was driving to a work site when a loud noise came from the opposite side of the Hunza river. A whole cliff was starting to collapse. They quickly stopped the vehicle and rapidly took a sequence of photographs as masses of glacial till fell into the Hunza river with a tremendous noise, causing water and dust to fly up into the air in a quite extraordinary way.

Surveying glacier snouts was also dangerous, as material was always coming away from the glaciers. It was Andrew's task to go around the base of the glacier snouts with the prisms on to which the surveyors aimed the distomat, and he frequently had to place the tripod very quickly, level the instrument, and then run away as fast as possible, because large stone boulders and fragments of ice were coming down all the time. It was clear that his chances of survival increased in inverse proportion to the time he spent under the ice snouts!

On one occasion, Andrew's cricketing abilities came into play when he actually caught a fragment in mid-trajectory. Similarly, David Jones was walking along with the plane table and tripod on his back one day when a boulder came down and hit the plane table, toppling David over and giving him a tremendous shock. The shock turned to horror when he saw that the tripod had been completely demolished and the Royal Geographical Society's plane table literally transformed into matchsticks. Had the boulder hit David directly, he would undoubtedly have been very severely injured or killed. On yet another occasion, Andrew was going up to the Gulmit Glacier, again with the prisms, by way of the moraine and by the side of a major melt-water stream coming down from the glacier. He was feeling rather tired and had his hands fully engaged with the equipment he was carrying, when he started to slip and felt the ground begin to give way under him. Immediately beneath him was a terrible surging melt-water stream in a deep rock gorge. He thought he was done for, but for some unaccountable reason, managed to stop just short of disaster. He then pulled himself back up to the ridge of the moraine. Since nobody had seen him, he then proceeded to descend on all fours towards his colleagues, who had to help him down since even on a very gentle slope he had completely lost confidence in his ability to stand upright.

Events such as these, which terrified even some of the most experienced members of the expedition, happened all the time. One day Professor Tahirkheli came into camp looking ashen-faced and bewildered. He was almost totally unrecognisable, the best clue to his identity being the geological hammer swinging by his side. He had been forced to abandon his vehicle and walk some 10km (6 miles) back to base after the highway had been blocked by a landslide. He had had to cross an extra-ordinarily steep scree, going down something like 900m (3,000ft) to the river beneath. This he managed to do by keeping his eyes closed and being piloted across the treacherous slope by porters.

Tahirkheli's hair-raising story brought back memories to Andrew, who that night dreamt that he was going across a steep scree composed of friable material, which was discharging into a great river. Gripping the bed hard, he suddenly shouted 'I can't go on, I can't go on in the dark,' and woke up with a pumping heart to find David Jones, still in his sleeping bag, but with knuckles white, holding onto the central pole of the tent. He too was having a nightmare on a similar theme.

Andrew reports that quite often at night one could hear people shouting, and that probably several members lived in fear that either they would fall, or that something would fall on them while they travelled the unstable landscape. Perhaps the most telling example of the hazards of the area was provided by the experience of two members of the Housing and Natural Hazards group at Gupis. Here a mudflow blocked the Ghizar river and formed a lake approximately 3 to 4 $km$ (1¾ to 2½ miles) in length. The flow not only caused the formation of this great lake but destroyed several wheat mills, drowned twenty-one houses, destroyed crops, killed two people and affected the livelihood of some hundreds more. The geomorphologists investigated this extraordinary phenomenon with some of the Housing Group, made a very detailed map of the mudflow and took samples of the material for analysis back in England. Another mudflow at Nilt, scene of some of the most heroic deeds of the British campaign to subdue Hunza in 1891, was also surveyed. Here the road was extensively blocked by streams of glutinous mud.

The way in which the Ghizar was blocked gives some indication of the dangers that such transient dams might create for the great dam at Tarbela on the Indus just above Attock. When the pressure behind a temporary dam gets to a critical level, the river breaches it and a great wave of water travels downstream for many kilometres, fed by the emptying reservoir of water previously formed by the dam. The Sarat landslide of 1857 blocked the Hunza river for six months and the waters thus compounded created an enormous lake, but suddenly, at the height of the summer flows, the dam gave way and a wall of water surged down the Hunza, into the Gilgit and thus into the Indus. Such was the volume of water that at Attock, some hundreds of kilometres away, where the Indus comes out from its gorges to meet the plains of the Punjab, the river rose by no less than 17$m$ (55$ft$) in seventeen and a half hours. An even more disastrous flood wave went down the Indus in 1841, when a landslide from Nanga Parbat dammed the river and created a lake that extended back for 55$km$ (34 miles). When that dam broke, the resulting wall of water caused a 24$m$ (80$ft$) rise in river-level at Attock. These landslips not only deeply affect people in the immediate area, but can also have repercussions for all those who live in the river gorges and for the inhabitants of the plains themselves. It is not inconceivable that a similar event today could cause permanent damage to Tarbela dam and even precipitate a domino effect.

It was to avert this danger that in 1974 the Pakistan Air Force bombed a mud flash which destroyed many houses in the village of Shishkat and then plunged across the Hunza river. It so happened that Assistant Professor Anis Abbasi of the University of the Punjab, Lahore, was inspecting the dam when the bombs fell into the soft mud around him. Fortunately the bombs did not detonate and he was able to beat a fast retreat.

This particular disaster caused large parts of the Karakoram Highway to be drowned by the newly-formed lake. Even when the dam broke it was too late to save several kilometres of the road, much of which was buried in new lake sediments leaving only the occasional parapet of a bridge sticking proudly out of the sand. A new section of road and a new bridge were quickly built below the dam, but both are threatened by the steep walls above, from which huge boulders frequently topple.

Another phenomenon with which the team was intimately concerned was that of rock weathering. All over the drier parts of the area, the rocks are covered in a dark brown, shiny material called desert varnish, which occurs on a wide range of rock types. Brian Whalley in particular collected large numbers of samples of this material in order to see how it is related to different rock types. As a complementary project, he established a whole series of high-altitude meteorological stations, up to 4,600m (15,000ft), in order to measure daily variations in rock temperatures and humidity levels and thus relate extremes of temperature at different altitudes to the processes of rock disintegration.

A further extraordinary feature in this desert area was the amount of salt in the landscape. Particularly when it was dry, white efflorescences of salt appeared in many places – including, unfortunately, the vicinity of some of the streams we used for drinking water. These efflorescences were probably composed of some rather nasty salts such as sodium sulphate and magnesium sulphate, and may well have been the cause of some of the diarrhœa which afflicted all members at base camp. These sodium sulphate and magnesium sulphate crystals also play a major part in rock disintegration, because on formation they expand in volume. A most striking example of how effective this process is in breaking up rock was shown by the state of the moraine deposited by the Minapin Glacier in the mid-1950s. In only twenty-five years, the rocks of this moraine, which were impregnated with sulphate salts, have been shattered in a most efficient manner.

The above findings have a direct, practical application in the Hunza Valley, where modern construction techniques are now being used. Local engineers must be made aware of the fact that there are substantial quantities of these very dangerous salts in the area, since their presence causes concrete buildings to decay rapidly. Already there are a number of signs that some of the bridges and other recent constructions along the Karakoram Highway are being attacked by these processes.

Much of the research done by the geomorphologists in the Karakoram was essentially academic, although obviously there were practical spin-offs in terms of where to build roads – and, more important, where *not* to build roads – how to protect modern structures, and how to locate particularly unstable areas, thereby avoiding catastrophes. In this latter context, some research was done on the frequency and distribution of the hazards on the Karakoram Highway, many of which are readily determinable from aerial photographs. It was a great pity these photographs were not made available to us sooner and for a longer period since, from detailed observations of slope angles and of rock-jointing and composition, the team were able to discern the various factors that cause major maintenance problems on the Karakoram Highway and assess alternative techniques for use by Pakistani engineers in coping with these continual hazards. Sometimes the maintenance engineers had to solve problems immediately with few resources, but they invariably upheld their motto: 'the impossible is done at once; miracles take a little longer.'

In conclusion, it is worth re-emphasising that here in Hunza nature creates a wide range of disastrous events which are almost impossible to describe in their extent and intensity in these few pages. It is surely a miracle that man continues to face these destructive forces and cling to his precarious existence.

# Living in a Hazardous Environment

On the way to visit the Prince of Hunza in his hilltop palace, I was stopped by a villager who was aware of the great influx of scientists into the valley. He asked if he could walk along the track with me and talk about our work. His English was flawless, and after many questions he said, 'I hope nobody drops an atom bomb here.' I was astounded. Perhaps he had learned that one of our Pakistani scientists was a geologist who specialised in uranium-rich ores, or that Pakistan was rumoured to be producing an atomic bomb. Why, I wondered, did these villagers, of all people, have to be concerned with that most evil of man's weapons? My feeble, but sincere reply was to state that men like him and me should try to convince everyone not to let such an event occur here or anywhere on earth.

Thinking over this encounter, I wondered if the villager had heard of the scheme to use nuclear power to melt the Karakoram glaciers, thereby providing a continuous all-year-round source of water for electrical power and irrigation. Whatever prompted his remark, it struck me as a most incongruous statement to make here in Hunza, where the villagers live under the constant threat of many natural disasters that can and do maim and kill them. The indigenous population of the Karakoram valleys has learned to live with these hazards and use their combined experience and skills to minimise their dangers, while accepting the inevitability of periodic catastrophe; they need and ask for no further complication to their lives. Since destruction is such a common event here, it was important that our scientists show their concern and provide every possible means of assistance to alleviate the problems of the local population and so illustrate that scientists can and do care about people, the environment, and all that is good about living. In this respect, the work programme directed by Ian Davis of Oxford Polytechnic had a great impact and was to have possibly far-reaching effects.

The most brutal of all forms of natural hazard is undoubtedly the earthquake. In Quetta, capital of Baluchistan, 30,000 people died in thirty seconds in 1935, the vast majority of deaths being caused by the collapse of inadequately-built brick buildings. Since then, great attention has been paid to the quality control of building materials, design specifications and methods of construction, and the Quetta brickwork pattern is now internationally recognised as being one of the most effective earthquake-resistant designs. In the Northern Areas, however, there are insufficient raw materials and no facilities for the mass-production of ceramic bricks. Homes have to be made of wood, boulders and mud, which in judicious combination can nonetheless resist the earthquake tremors that all too frequently occur.

In the village of Patan, now bisected by the Karakoram Highway, a major earthquake in 1974 was reputed to have killed 2,000 people (although actual deaths may only have been 800), many of them caught in collapsing buildings. It was this disaster which led us to research into possible improvements in building design. A superficial analysis of the structures indicated that the villagers had built them with earthquakes in mind, since the heavy earthen roofs were independently supported on timber columns in lieu of side walls. Furthermore, some horizontal and vertical vibration of side walls could be accommodated by the ingenious construction, since this included lateral and upright timber members in each wall, permitting far more sway than in the more modern, European type of building. One objective of our study was therefore to examine the thesis that the indigenous population had learned, possibly through centuries of experience, how to build their homes in an earthquake environment. If we could learn from their experience, it might be possible to assist other developing countries that suffer similar hardships.

The group selected by Ian Davis to study the question of living with natural hazards included engineers and scientists of many disciplines, all of whom could individually make a valuable contribution to the project, not only with respect to work on the vulnerability of housing, but also in connection with the study of other natural hazards of the Karakoram, such as floods, fires, drought, erosion, rockfalls, landslips and disease. It thus came about that this group became an interdisciplinary team in its own right, and one that was flexible enough to rearrange its priorities as the research work advanced and the group became aware of the different order of priorities in the problems faced by each village.

With the assistance of Professor Nicholas Ambraseys of Imperial College, who acted as an earthquake engineering consultant, a programme of work slowly evolved. Ian, a professional architect, has gained an international reputation for his work related to hazardous-living environments, and initially it was thought that he and his group would have little difficulty in raising funds for what was obviously a first-class programme that would have a direct impact on improving the lot of mankind.

Dr Robin Spence had devoted much time and effort introducing technical innovations to poorly provisioned villages in developing countries such as Nigeria, where I also had worked for several years. Professor Cliff Moughtin from Nottingham University, an ex-Nigeria colleague, was also experienced in this type of work especially since he too had many links with developing countries through his work as a community planner. I had no doubt that both these men would be energetic and conscientious workers in several areas of importance to us: Robin was a structural engineer with experience of low-income housing with particular emphasis on low-cost building materials; Cliff was an architect with experience of taking into account the religious, cultural and social needs of a community, combined with knowledge of how to make optimal use of the available space and the conditions of the local environment.

Another member to join the team was Dr David Nash of Bristol University, a colleague and friend of Robin's with specialist interests in structural mechanics and seismology. This youthful enthusiast carried around amongst his tools a device which, when clamped to a house, would cause it to vibrate. The whole system was designed and manufactured by Robin and David on a Science and Engineering

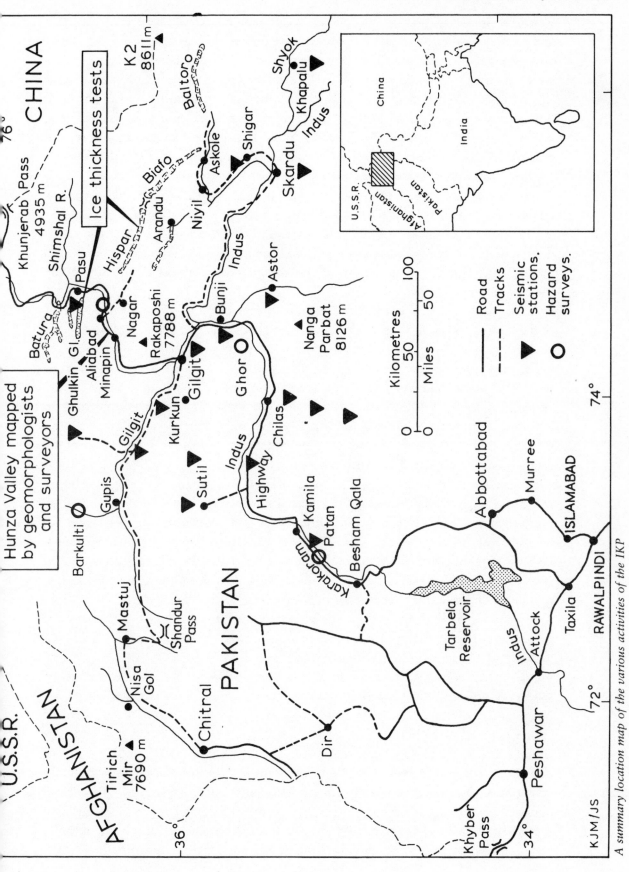

A summary location map of the various activities of the IKP

Research Council grant, and it included accelerometers, amplifiers, an oscilloscope and a tape recorder to enable the vibration to be not only seen, felt and heard, but also permanently recorded. Fortunately David's charm and perpetual warm smile were sufficient to allay the fears of any distraught house-owner who found himself shaken out of bed. David's major task was to make a detailed examination of buildings in the area; he would then design and construct models based on them, which could be tested on his vibration table back home at his university and evaluated for their resistance to shock. In this way it was hoped to formulate designs for the best shock-resistant structures made of local materials.

Another self-effacing young member of the group was Richard Hughes, who was qualified both as a geologist and an archaeologist. He had had much experience in the restoration of historic buildings, particularly in the Middle East, where his double skills had been employed in analysing the performance of building materials and their deterioration characteristics, and mapping the various hazards suffered by the community.

At the last moment, the team was fortunate to acquire the services of the Director of the International Disaster Institute, Frances D'Souza, although unfortunately she could work with us only for a limited period. She is possibly the only anthropologist in the world to specialise in human behaviour patterns related to disasters. Her work was of the highest quality and was crucial in the study of Barkulti, a very isolated community in the Yasin Valley (see Figure, p. 115). Frances was able to gain access to homes to elicit vital information from the local womenfolk on their priorities and perceptions of risk, an impossible study for the male members of the group in such a strict Islamic society.

Other important contributors to the work of this group were the expedition doctors. They not only recorded the physical illnesses and mental stress created by the many hazards of the Karakoram, but also accumulated a vast store of goodwill without which many doors would literally not have opened. This was most important in respect of contact with the womenfolk, who were ably treated by Helen Massil. As a result of her medical work, much valuable information was collected, in particular that relating to the nutritional problems of the community.

Fortunately the expedition also had the services of Professor Israr-ud-Din, a social geographer from Peshawar University. Israr, a tribesman from Chitral, had studied at King's College, London, where he had completed a thesis on housing in the north-western corner of Pakistan. Quiet, modest, but also tall and authoritative, he seemed completely at peace with himself and his surroundings at all times. He was to be of great assistance – and not only because of his long and deep personal experience of the neighbourhood. Israr also introduced to the group a second-year geology student called Shahbaz, an extraordinary piece of good judgement on his part, since this young man was able to invite the group into his home in Barkulti, where his father was a prominent man in the community. As a result, virtually all doors in the village were open to us and answers to many questions translated rapidly from the local language into English. Information on a wide variety of topics was gathered in this way, ranging from social patterns and priorities to Cliff's

Khurtlot Village
Soharabad. Pakistan.

*Left* A house in
Khurtlot village
painted in water-
colour (*above*) by
Ian Davis. Note
horizontal timbers
giving lateral
flexibility.

Apricots drying in
Hunza.

Building a hay stack
in Hunza.

queries on land usage, ownership problems and settlement development. This piece of good fortune, coupled with the fact that there were three female members in this team (Shane also assisted the work of this group), meant that a most comprehensive study of Barkulti was able to be completed. Without our female membership, it would undoubtedly have been impossible to gain access to the innermost parts of houses. Only in the Ismaili communities at Aliabad and a few other villages of the Hunza Valley was ease of access comparable, but this was because the Muslim sect to be found there were more open, and also because of a long period of care and affection from our two hardworking doctors.

Yet another fortunate addition to this team was Dieter Illi, an architect from Switzerland, who for the past seven years had been researching into the derivation of house forms in the Northern Areas of Pakistan with Professor Carl Jettmar of Heidelberg. Dieter assisted in the task of understanding the precise purpose of each part of a house and in establishing their age and different characteristics. He was a first-class draughtsman and during the course of the expedition produced several very beautiful drawings, as did Ian Davis himself (see p. 116). Ian's watercolours attracted many onlookers, who soon became friends and subsequently helpful assistants to the expedition.

Thus, Ian brought together a team of experts from a number of disciplines and was able to direct a comprehensive programme that was complementary to his experience in Central and South America, the Middle East, India and disaster areas of Europe. This time, however, in contrast to his previous work, his team were looking at means of reducing the vulnerability of the population, rather than at the problems of reconstruction after a disaster had occurred. It was to my amazement and regret, therefore, that the group did not manage to persuade the Social Sciences Research Council to make some contribution to the costs of the programme, one that I and many others considered to be a most valuable and humane piece of research with long-term international implications. In fact, this group had more difficulty in raising funds than any other programme, which was the very opposite of what we had expected. However, in the last few days before my departure I had drafted a very detailed proposal on all our programmes to the British government, which was up-dated daily by letter and supplemented by additional information given by telephone in answer to urgent queries as the proposal went from desk to desk. The nail-biting exercise paid off, since on the day I left for Pakistan a grant was awarded that permitted us to assist each and every one of the expedition projects. Thus on 23 July, soon after completing his duties at Oxford, Ian was able to come to Pakistan. Fortunately Robin and the team had initiated much of the fieldwork and so there was minimal inconvenience and delay.

The projects of this group had involved more discussions, travel, meetings and planning than any other programme. It was almost certainly the first time that any single multidisciplinary group had examined the problems of living in an area of intensive and extensive natural hazards, either in Pakistan or anywhere else in the world, but at the outset no one could have anticipated the scope and depth of research that would emerge from such a heterogeneous body of scientists. Unlike the other programmes, this group had to establish a research methodology, since there were no precedents to work from.

It soon became evident that it was possible, indeed necessary, to modify existing housing styles so as to reduce their vulnerability to earth tremors, but such changes could only be implemented in new constructions. However, the most important finding concerned the indigenous population's priority rating of risks. The most crucial, day-to-day problem relates to the availability of water, or rather the lack of it. Severe drought – the annual rainfall in the area is of the order of only 7cm (3in) per year – is often matched by severe flooding, caused by heavy localised rainfall, rockfalls, landslips, glacier surges or mudflows. In winter, when the sun no longer melts the glaciers and snows, there may be no water for the village; a situation that is very difficult to imagine during the summer months when water is so plentiful. In Hunza, one large village is now completely deserted: there are no children in the streets, no shy women disappearing into the labyrinths of the houses, no smoke drifting from holes in the roofs, no grass, indeed no vegetation of any form; just a high, levelled platform on the precipitous flanks of the gorge, containing a few crumbling walls – and all because the melt-water stream from a glacier hundreds of metres above changed its course and caused flooding in a neighbouring village. Little wonder that many years are spent in the planning and construction of new villages, and that aqueducts many kilometres long are built across vertical cliffs so as to ensure a safe and long-lasting water supply. The only mitigating factor is that the population is highly concentrated in their villages; as a result, in statistical terms, the numerous hazards they face are minimised, although one side-effect is that the outside world takes less note of the recurring disasters than they should. However, there are also some more insidious problems. As each village extends, land suitable for cultivation becomes more difficult to find, and meanwhile the steady erosion of what there is continues unabated. Fields on the edges of collapsing precipices above the river gradually disappear, as do the fields at the rear of the village which are the catchment areas for rockfalls and landslides from the cliffs above. Facts such as these became apparent as each scientist presented his daily findings during informal discussions at the end of the day, often extending well beyond each evening meal.

During studies in the village of Barkulti, the team was told that, in the event of an unusually long spell of rain (by far the most serious threat), one villager would be posted on duty all night to listen for rumblings on the hillsides which might herald a land- or mudslide. Unfortunately, when such catastrophes do occur, houses are often sited in the worst possible situations, as evidenced in 1974 when half the houses in a Hunza village below the Balt Bare Glacier were destroyed by a mudslide. The reason for building homes on such dangerous sites is that the people are living at a very low subsistence level, and their survival depends on all land suitable for cultivation being developed, however meagre. In consequence, the steepest of hillsides are made into fields by constructing massive flights of steps with high vertical retaining walls, behind which are laid boulders, then stones, then rubble, then soil. The retaining walls may be as high as 3m (10ft), but the 'fields' only 1 to 2m (3 to 6ft) wide. The result is that a small, steep cliff face is transformed into a series of narrow fields that have the appearance of a giant's step-ladder, or, when being irrigated, a cascading waterfall. Houses are built on what little land remains, frequently below exceedingly steep precipices, or worse still, on the flanks of gullies that carry glacial water; hence the threat of surges of melt-water from far above in the mountains that can rapidly turn into pulsating mudflows.

A further problem is posed by the growth of the local population, which has led to the development of more fields in increasingly inhospitable areas, thereby increasing their vulnerability. Such is the case in the villages around Yasin, Hunza and Nagar, but as yet there does not appear to be an overall strategy to overcome common problems, and each village is having to find its own solutions.

However, rather than moving houses to flat sites on the most valuable fields of a village, it is more practical to implement improvements in the basic economic and social structure of the village, including provision of educational and medical facilities. Few improvements can be made in husbandry, and indeed the villagers do outstrip us in their attention to farm productivity and the care of livestock. Goats, sheep, hens and calves live in the outer sections of houses and are treated almost as pets. This enables the supply of feed, which is always meagre, to be closely controlled. Free grazing is never allowed, and children keep a careful watch on all animals. It is not uncommon for three crops per field to be grown each season, and the earth is regularly fed with human and animal excrement. The stalks of corn spread over the land like a thick mat right to the very edge of each field, whilst the tops of some of the back retaining walls, though only a few centimetres wide, are used as village paths as well as irrigation channels.

My own discussions with the Prince of Hunza, Ghazanfar Ali Khan, centred on these and other related problems. He wishes to create more schools and improve the medical facilities available, but now that Gilgit is more readily accessible, the young men are being drawn away from home, and this creates many labour problems. For example, it used to be customary for young boys to be responsible for rising early in the morning and changing the course of the water-flow in the irrigation channels. Starting at the top of the village, they would remove mud and stone plugs and use this material to seal off the old channel. Each stream is allowed to flow for a period of time determined in council by the village elders, after which the flow is diverted to another field. Some fields are only allocated a little water each day; some are watered only once per week or even per month, as was the case with our base camp area, an orchard of apple and apricot trees. Today, the work of the boys is often being done by old men, and it is difficult to see how the village can continue to survive the rapidly-changing social patterns. The older inhabitants are also exposed to greater dangers now, since it is often necessary for them to leave their distant but safely placed houses in order to live closer to the fields, which need to be tended continually.

The educational priorities are to train engineers, geologists, lawyers and doctors. It is recognised that engineers can improve the design and construction of buildings, provide and maintain electricity generation plant, and improve the irrigation systems. For example, the present fifty-year-old second-hand water turbine at

Hasanabad in Hunza provides less than one-thousandth of the power of a modern European turbine, and then only spasmodically. I spent a day helping to strip down the turbine alternator bearing, only to find that the bearing metal was badly worn, so helping to excite excessive vibration that was primarily due to misalignment of the shafts. What was really needed was a new turbine, but instead we had to make do with smoothing the bearing metal inside the now badly worn cap and ordering another bearing. Perhaps such continual repairs will last until the glacier above surges forward once again, devouring the turbine house and the huge aqueduct that feeds it.

As more water is required by the expanding villages, so more lawyers are needed to arbitrate in the ever-increasing controversies concerning the supply lines and distribution of that vital commodity. One consequence of increasing legal involvement is that the lawyers are also introducing the new laws of Pakistan which impinge on the daily lives of the population. Hunza was a self-governing kingdom for 600 years until Bhutto, in an effort to curry favour with the underprivileged, banished the Mir (king) and ended the feudal system of government. Anarchy and corruption grew, and in this sad period far too many trees were felled by the new peasant landowners, who of course now realise their folly and see the need to maintain a balanced ecology. Under the old regime no one could remove a single tree without permission of the village council headed by the Mir, whose decisions were governed by a strategy based on hundreds of years of experience.

As the new political scene develops in Pakistan, Hunza is attempting to utilise the best of both systems. The Prince, still referred to locally by his friends as the Mir, is regarded as the leader of the villages of Hunza. He maintains his Palace in Hunza and vacation homes in Ghulkin and in Islamabad, where he sits as the elected unopposed member in the ruling Council. He and his beautiful wife are held in great respect by the people, but much of the day-to-day government of the valley is in the hands of the District Commissioner, who controls legislation in the area through the local police force. The Pakistan Army is also represented through the Frontier Works Organisation which, as we have seen, is responsible for the daily maintenance of the Karakoram Highway.

There is also an urgent need for trained geologists to work in the area, since they are the scientists best equipped to explore the mountain ranges for mineral ore and precious stones. There are several low-grade ruby mines in the area, and gold panning still goes on in most of the beds of the Hunza Valley rivers. No one has yet found a rich source of any mineral, but one cannot help thinking that an area so devastating as this should have some compensating feature somewhere. A greater understanding of rock structures and composition is also required to assist road and village construction, as well as to identify particular zones of weakness in the mountain walls and thus minimise the dangers which continually threaten all forms of human activity in the valleys.

The fourth and most important priority in the development of the area is the provision of medical facilities and trained personnel, but, as is common elsewhere in the developing world, once students qualify they tend to seek more lucrative practices in the big towns and cities of Pakistan, the Middle East, or the western world. As yet, there is no resident doctor in Hunza to tend a people starved of basic care and attention, and for three months our doctors answered each and every call made on them, working at all hours of the day as occasion demanded.

It is almost inevitable that the brain-drain will continue unabated, and new schemes will have to be introduced to combat the growing problems of these remote areas. Certainly it would bring no satisfaction to those of us who have been privileged to clamber up the tracks of these lonely gorges to see them continue unchecked. We were therefore very glad to be visited by a representative of the UK Overseas Development Administration. It was his opinion, later to be reinforced by other officials, that our work was more cost-effective than other exploratory and development schemes for overseas aid; in his view we were getting to the heart of many problems, putting experts into direct contact with the people requiring assistance and by-passing the lengthy bureaucratic and inter-governmental channels. Our own Housing and Natural Hazards group was viewed as an intensive, highly expert unit, and furthermore, because of its interdisciplinary nature, it was felt that its members were able to provide a uniquely coherent view of the real problems involved and suggest optimal solutions.

As an example of the need for work of this kind, it is worth mentioning the studies at Patan. In December, 1974, a 6.0 magnitude earthquake destroyed the village and it was estimated that 800 people died. It took several days before the gravity of the disaster percolated to the west and the aid belatedly poured in. Because the tragedy occurred in the middle of winter, a huge and rapid reconstruction programme was required to provide warmth and shelter for the hundreds of homeless people, but unfortunately there was no one on hand to provide immediate guidance to the builders. As a consequence, many of the new houses are far more vulnerable to tremors than their predecessors. It was apparent to us that a new style of construction was now infiltrating the valleys. In contrast to the traditional buildings, with their masonry walls, timber strapping around the edges and earthen roofs laid on top of timber beams, the new structures are composed of stone, brick or concrete-block walls, normally covered with a lightweight corrugated-iron roof. One of the reasons for this changing pattern of construction may be the influence of what is considered to be the 'modern style', but certainly the cost of local wood and the need to preserve the dwindling supply are also important, if not paramount considerations. Sadly, extensive deforestation has taken place since the building of the Karakoram Highway and, as in other parts of the world, an immediate review of the situation should be instigated. Failing this, the villagers will have a man-made catastrophe to add to the already immeasurable natural hazards they face.

It is doubly sad to report therefore that many of these new houses were poorly constructed. Unfortunately, local builders did not take into account the fact that, when new styles and materials are introduced, it is not only necessary to employ them to best advantage, but also to be aware of the benefits as well as the faults of previous methods and materials, if only to make sure that simple mistakes are avoided. Wood, stone, mud, concrete, steel, and bricks all have both advantages and disadvantages and should be assessed both individually and in combination for their usefulness and limitations. Wood is a composite material that will flex and can take compressive loads, but it cannot take high-tensile loads, which cause it to splinter. It is an ideal structural, framing material, but supplies of it are limited and it also burns fiercely. Stone does not burn and is a good insulator, keeping heat in and cold out; it can withstand compressive loads, but easily cracks when under tension. Mud is also unable to withstand tension, but it is an easily moulded bonding material which is not as heavy or expensive as concrete and is suitable for roofs and non-load-bearing walls. Steel has to be imported, but stones are naturally available in abundance. Taking the cost and properties of all these materials into account, it should be possible to devise an effective and economic type of building that will help reduce the number of deaths caused by buildings collapsing in earthquakes.

Reinforced concrete has revolutionised the construction of buildings, but it does need to be used properly, with the steel correctly positioned in the concrete and with the ends of the reinforced beams firmly attached. Where concrete was available in Patan it was often used inappropriately, with no understanding of its uses and limitations. Sometimes unattached steel bars could be seen poking out into space at the ends of concrete beams, and in other instances the tensile steel bars of the reinforced beams had been positioned at the *top* compression side of a beam instead of at the more fracture-prone, tensile-stressed bottom side. In another village, the concrete posts between shops had been chiselled away to reveal the steel reinforcement, onto which door hinges and shutters had been welded.

From these simple construction faults it was clear that there is a major need to train local builders in the arts of modern construction using new materials and to instruct them on simple safety requirements, pointing out what damage an earthquake is likely to cause. Although the need for training has been clearly identified, the problem has been compounded by the fact that the new breed of craftsman is frequently induced by higher wages to work in the towns and cities of Pakistan or to migrate to the Middle East. Nevertheless these problems have to be faced and solutions found if catastrophes such as occurred in Patan are not to be repeated.

Finally, it must not be forgotten that many high-altitude alpine villages are composed of all-wood houses; the greatest threat during an earthquake is therefore that of fire. What makes this hazard far more serious is that the Karakoram is a desert, and even in the low-level inhabited areas water is scarce and not immediately available to douse the flames as they quickly spread through the dry timbers. In the summer of 1980, one village of 240 wooden houses was completely destroyed in this way.

A further problem concerns the health risk to outsiders working in the villages. Many of the housing group's studies were concentrated within village communities, where members were subjected to contact with diseases against which they had no natural resistance. All members became plagued by house fleas, which caused intense discomfort and annoyance; it seemed to make no difference how often people plunged into river waters or how long they soaked themselves, the fleas always survived better than their hosts, who were eventually forced to climb out of the icy water, teeth chattering. The villages' water-supplies carried yet another irritant in the form of suspended particles of mica, which act as microscopic razors and cut the inside of the stomach, causing various intestinal complaints. It is almost impossible to avoid these various ailments during a first visit to the high mountains, but fortunately the effects decrease and usually disappear on subsequent expeditions.

The most dangerous hazards in the area, however, arise from the perpetual war fought between the mountains, rivers and glaciers, the ammunition being supplied by snow, rain and wind. One incident that occurred near Gupis on the Ghizar river near the lower end of the Yasin Valley could have caused the death of both David Nash and Richard Hughes. A pulsating wall of mud swept down the gorge behind the village during the evening of 26 July. Each pulse interval was timed at fifteen minutes and was 8*m* (26*ft*) high. As each wave passed by, the sound was deafening, not unlike an old steam train rushing by a few centimetres from one's ear. Discharging into the main valley, this fast-moving composite river of grit, stones, boulders and water added soil, bushes, walls and houses to its sticky mass as it flowed relentlessly into the main valley, blocking the course of the main river (which is the size of the Thames) and immediately creating a lake that continued to expand during the remainder of the expedition. The new lake and dam will eventually collapse, threatening the bridges of the Gilgit and Indus rivers and ultimately the dam at Tarbela.

Perhaps somewhat foolishly, David and Richard walked into the mud-flow, which had the consistency of quicksand, to take measurements of the height of the pulses and their periodicity. Fortunately they escaped when the situation rapidly worsened, but two villagers were less fortunate and were pulled into the flow and drowned; several houses in the village were also destroyed. Richard and David later monitored the rise in water-level upstream of the dam, which was found to be 0.5*m* (20*in*) per hour. On talking to the families whose homes had been destroyed, Frances D'Souza found that, rather than receive succour and shelter from relatives living elsewhere in the valley, they preferred to stay and display a suitably destitute appearance in order to qualify for aid from the District Commissioner's office in Gilgit.

After Barkulti, the group's next detailed case study was centred on the earthquake village of Patan and included an analysis of property that had been destroyed and not rebuilt, new buildings, damaged but modified buildings and finally old but undamaged property. By the time they had completed this study they were well-versed in the techniques of gathering data, had become accustomed to the problems and were able to tolerate all the various illnesses associated with this area.

From Patan, the team moved back up the Indus to Ghor for a third study. This village is situated at about 3,000*m* (10,000*ft*) and looks across the great river gorge to the precipices of Nanga Parbat, ranging some 7,000*m* (23,000*ft*) above the valley floor (see Figure 13, page 115). The village of Ghor is situated just below the mountainside tree-line and fortunately timber was plentiful and readily accessible here. The architecture of the village therefore relies heavily on timber for the main frames of buildings, with stones providing the infill.

The group had a cool reception here, because the villagers suspected that the intense burst of photographic activity was not related to an interest in their house construction styles, but was aimed surreptitiously at their women. Such misconceptions can lead to nasty incidents, but fortunately the women members of the expedition managed to restore calm and eventually the team were permitted entry into some of the most exquisitely styled homes yet seen. Ghor is a delightful village, and much was learned about its social and physical structure, how houses were built and maintained, and the various uses both of space and individual items within the houses. The expedition group also took over the village dispensary, which, as in many villages in the Karakoram, is seldom stocked and used. The inquisitiveness of the children was matched by that of their parents, many of whom had never seen white faces before, and the team's every action was carefully watched and noted. Eating, talking, working and even sleeping was monitored to see if we foreigners had any unusual characteristics. Translation of the local dialects was simplified by enlisting the assistance of the local schoolmaster who rendered the expedition an invaluable service. Villages such as Ghor are very remote and do little trading and so have to be self-sustaining. Without the schoolmaster's assistance it would have been difficult for us to communicate with the villagers.

The fourth and final study was conducted in Hunza itself, at the village of Ganesh. Here, Karim, Manager of the Tourist Cottage in Gilgit, was the catalyst for the early exchange of data. At his own expense he took time off work and travelled up to Hunza to introduce members to his relatives, teachers and friends. The village, on the right bank of the Hunza river, is the only outpost of a strict Shia faith of the Muslim religion among the more tolerant and liberal Ismaili communities of the Hunza Valley. Unfortunately, matters got off to a poor start when the film crew descended on the village unannounced, unaware that the villagers had different patterns of behaviour to the inhabitants of other villages in the immediate vicinity, especially regarding the photography of women and children. However, there were an adequate number of other studies still to be done in the vicinity to compensate for this setback. Visits were made to the historic fort of Baltit that had once been the stronghold of the Hunzakuts. It is a remarkable structure. Built at the top of the village, it lies under the precipices that soar up to the summits of the Batura Muztagh and stands on its own high mound with steep cliffs on three sides, an entrance being gained by a steep but wide stairway up the fourth and front wall. Inside are many historical relics, including chain mail armour. Photographs of British royalty are displayed in one of the main rooms, but unfortunately both they and the rooms are in a sorry state of repair. The small hole for prisoners vividly portrays the way of life that existed here until 1891, when Durand broke the rule of the Mir and stopped the Hunza peoples' raids for plunder and captives. Richard writes as follows:

'The fort proved to be of great interest to the housing group and had withstood an earthquake, surviving because the construction technique employed allowed the walls to flex much like human vertebrae. It is generally believed that the fort was built some 600 years ago, when a Princess of Baltistan married the reigning Mir of Hunza. The princess's father is reputed to have sent with the bride an army of Balti masons, carpenters, and craftsmen who built the fort, and also Altit Fort, as part of the royal dowry. Certainly its construction techniques are reminiscent of Baltistan, but from my survey at least seven phases of construction were recognised and it is more sensible to assume that the fort slowly developed into its present form. However, the fort is by far the oldest thing in the Hunza Valley with the exception of the mountains themselves.

'We found many signs of serious decay, particularly in the roof and in the rear walls. In Baltit Fort, as in all the houses in the valley, extensive use has been made of the local micaceous soil for the mortar between stones and for the roof covering. Where such material is used there has to be constant maintenance. The Mir moved to his present palace in the 1950s, leaving the fort to fend for itself against the elements. Unless repair work is carried out in the near future, the roof and rear walls will become unsafe and the whole building will enter its final stage of rapid decay.'

Hopefully, Richard's full report on Baltit Fort, based on his thorough investigation, will incite the appropriate authorities to fund a restoration programme, since it is probably the most historic of all buildings in the entire range of the Karakoram mountains and as significant to the Northern Areas as the Tower of London is to Britons. A survey was also made of the much smaller Altit Fort, which also stands on very high ground with one precipice descending directly to the Hunza river. There is little doubt that it was from here that unfortunate villagers convicted of crimes had to throw themselves to their death at the command of the Mir. Banishment to the remote village of Shimshal was considered a similar fate, but at least there they could atone for their crimes by providing food and shelter for the passing parties of brigands under the Mir's control.

In the final analysis, the Housing and Natural Hazards programme achieved far more than was intended or thought possible after Ian's early reconnaissance trip to Pakistan in February. Although this had yielded some dividends, he had found it impossible to solicit all the assistance required or to forge strong links with the people with whom he would eventually work. Nevertheless, in the event, at least a dozen local inhabitants, headed by Riaz, manager of the Chinar Hotel at Gilgit, rendered the project invaluable service, and when the time came it was hard to say goodbye to such good friends. This was also true of the surveyor Mohamed Farooq, who had mapped the layout of Barkulti on our behalf.

Before returning home, Ian and Robin gave a series of lectures in Lahore, and Ian repeated the exercise in Karachi. Much detailed analysis is now being conducted, and a wide network of interested parties is involved in discussions on future programmes. In particular, it is hoped that the UNDP will accept a simplified development plan for the Yasin Valley based on the data derived by our group. This and similar sets of comprehensive data related to the work of this and all other groups were presented at the RGS in September, 1981 (see Appendix IV, p. 199).

In conclusion, it should be noted that this group probably learned more than any of the others who, unlike them, employed established techniques for gathering data and used modern, sophisticated equipment. The other groups had been able to go about their tasks immediately on arrival at camp, but the Housing and Natural Hazards group had slowly to learn new techniques, gain the confidence of the villagers, single out the problems of most importance and then implement the most appropriate means to complete their study. The fact that they completed their work without a serious accident and left behind so much goodwill is in itself sufficient reward. I do not doubt that the next party to visit these remote and lovely villages will have much for which to thank their predecessors; but no doubt they too will be shocked to learn that, in these low-subsistence villages, the child mortality rate can be as high as 70 per cent and that in 1980 eleven people were killed whilst traversing the tracks between the villages beyond Gilgit. They too will wonder how and why these peoples, the greatest survivors in the world, continue to cope with the galaxy of hazards which daily threaten their fragile hold on the landscape. Wedged between four nuclear nations, perhaps they can continue to teach us how to survive.

# Into the Ice

On 27 July I returned to Gilgit from Islamabad, where I had been liaising with the British Embassy staff in attempting to co-ordinate the release of the sad news of Jim's death. Back in Hunza there were several problems to solve, the most pressing of which concerned the glaciological unit. At the outset this team was composed of eight technical members, Gordon Oswald (Director), Marcus Francis, Professor Zhang Xiangsong, Dong Zhi Bin, George Musil, plus Jim Bishop, Ron Ferrari and myself. Jim, Ron and I had worked together on the discovery of under-ice volcanoes on the Vatnajökull ice-cap of Iceland, where we had used radar equipment similar to that which we had with us now to determine the thickness of Karakoram glaciers.

We knew that Ron could only stay for a matter of days, but he would at least be able to help assemble the instruments. Likewise, Gordon Oswald was only permitted to stay in Pakistan until mid-August, when he was due to return to the UK before leaving for China as a technical representative of his company. To compound these difficulties, we now learned on our radio link that George Musil was very weak and needed medical attention, and it was later confirmed that he was suffering from hepatitis. We also heard that Xiangsong had slipped on the glacier and had badly sprained his ankle. Later we were to find it was broken. Thus, soon after Jim's tragic accident, the team of eight was reduced to three: two electronic engineers, Marcus and Zhi Bin (now affectionately called Lao Dong), and myself acting as a surveyor. Fortunately, the team included the film crew of Paul Nunn, Tony Riley and Ron Charlesworth, who were all excellent mountaineers and were providing route-finding support in order to get our two radar units quickly deployed on the Hispar Glacier.

It was essential to review the situation on the spot, and anyway, I wanted to get away from the administrative chores and hassles as soon as possible; Nigel and Shane were more expert than myself at handling these day-to-day problems, and I also wanted to make some technical contribution to the programme, however small. One further problem concerned Major Rana who, although attempting to disguise his curiosity about what we were doing in the Hispar Valley, was very nervous about our activities. He was officially required to have first-hand knowledge of the team's work, but he was no mountaineer, and by now several detailed and ungarnished stories of the dangers to be faced in the Hispar Valley had been recounted by an obviously sincere Ron Ferrari, in somewhat understated terms typical of British academics. These, of course, had the contrary effect of emphasising the dangers to the uninitiated Major. Meanwhile we kitted him out and I taught him the simple climbing-rope knots and the use of prussik loops for climbing out of crevasses. Bob Holmes, the official photographer of the expedition, also wanted to join our team,

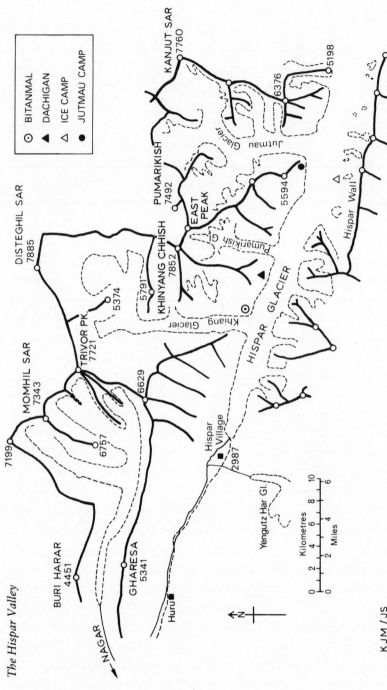

*The Hispar Valley*

and of course David Giles was to accompany us to give medical advice to our sick companions. We planned to be away about three or four weeks.

The first party (Gordon, Marcus, Professor Zhang, Lao Dong, George, Ron Ferrari, Paul, Tony, Ron Charlesworth) had left Aliabad more than three weeks previously, taking with them nearly 1,500*kg* (1½ tons) of supplies and equipment. The intention of this party was to work on the Hispar Glacier for the duration of the Project, as an expedition in its own right. The glacier is over 50*km* (31 miles) long, rising from 3,500*m* (11,500*ft*) above sea-level to 5,200*m* (17,000*ft*) at the Hispar Pass. It presented an opportunity to sound the depth of ice over a widely varying set of conditions, including different ice temperatures, changes in glacier cover (rough and smooth, ice and moraine), and variable thickness both in longitudinal and transverse profiles. The technique was to use electromagnetic radiation (radio waves), which travel through ice and are reflected back from the glacier bedrock to a receiver. Using the fact that the speed of radio waves in ice is 169*m* (554*ft*) per microsecond (a millionth of a second), it is possible to measure the thickness of the ice from the time interval between transmission and reception of signals. By moving the equipment over the ice, manually or in an aircraft, and plugging the receiver into a synchronized ciné camera, it is possible to record the depth and profile of the ice automatically, and so determine the three-dimensional shape of a glacier – something that had never been done on any of the glaciers in the Karakoram or Himalaya. Before full use could be made of this technique, however, several problems had to be resolved.

Because much of the Karakoram ice is at or about melting temperature, the electromagnetic radiation is absorbed before it can be recorded in a receiver. A further problem is posed by the presence of much water within the glacier itself, which has the effect of scattering the waves, thereby dissipating the strength of the signal. Knowing this, we had brought with us a new form of radar equipment, called impulse radar, that transmits 15,000 high-energy pulses per second, each of average duration 0.5 microseconds. This equipment had proved very successful on the very wet Vatnajökull ice-cap in Iceland in 1976 and 1977. However, here in the Karakoram we were faced with the potentially complicated problem of ice in a deep valley, where echoes from the side walls could be mistaken for signals from the bedrock.

To make certain that we acquired some results and obtained fundamental data on the electrical properties of Karakoram ice to assist further development of this kind of equipment, we took two sets of impulse radar equipment to the Hispar Glacier, the 'old' and the 'new'. The old equipment was of the Vatnajökull type, which had sounded depths of 900*m* (3,000*ft*). Unfortunately, this equipment had very long antennae (120*m*, 394*ft*), which would make traversing the glacier difficult. The new equipment was more suitable for penetrating cold ice; it was more compact, and was built to make use of digital signal-enhancement techniques and digital magnetic-tape recording devices. Unfortunately the new equipment was as yet untried but we hoped that either one or both sets would prove satisfactory – and there was always the possibility of modifying one set with components from the other if the need arose.

One further aspect of this work was the necessity to survey the glacier, so as to relate the points and traverses of ice-depth sounding experiments to the local topography. It would be pointless determining the depth of ice, if the exact positions of the soundings were not known. Due to Jim's accident, no trained surveyor was with the glaciology team and with Xiangsong and George very poorly it was essential that we reinforced the Hispar party as soon as possible.

The practical relevance of this work was to determine the storage capacity of such glaciers, as they are in effect frozen reservoirs. From three-dimensional diagrams of these great solid rivers it would also be possible for us to assess the potential energy of the glaciers and the risk of surging. But first our instruments had to be tuned and systems devised to allow the equipment to be moved rapidly over the glacier surface, hopefully by helicopters. The Hispar Glacier was to be an 'academic' exercise to see if the equipment would work. If and when it did, we could then start on measurements of ice thickness over as large an area as time and weather allowed.

Back in Hunza I spent a couple of days repairing kit, collecting money to pay our porters, packing equipment, visiting the Commissioner in Gilgit to pass on our frustrations about our lack of passes to travel to the Chinese border (we now had our Chinese visas), and to visit Hayward's grave. On the evening before my departure, I addressed all members currently at base to inform them of the state of the game, with particular reference both to our unwanted political problems and the need to extract as much scientific data as was possible in the remaining period. Finally, I told everyone of the arrangements agreed in Islamabad for greeting our distinguished visitors, Lord and Lady Hunt. We intended that as soon as they arrived at base camp we should assist them to plan a memorable climbing holiday to suit their own requirements.

On that last evening I was thankful to be sneaking away for a short while, and I took the opportunity of plucking a couple of apples and a far greater quantity of apricots off the trees in our base camp. It would be a little time before I could repeat that pleasure. Before turning in, I looked across the hillsides and absorbed the gold, brown, yellow and deep rust colours of the great scree slopes above Aliabad and the magnificence of Mount Rakaposhi. The final delight was knowing that soon I was to leave behind the great hordes of black flies that immediately swarmed over the tiniest morsel of food.

Bob and I left in the early hours of 29 July aboard a jeep stacked high with equipment (Rana and David left later). On the far side of the Chinese bridge spanning the Hunza river I surveyed the bleak stony hillside where irrigation channels were to be built prior to the reclamation of land for another 2,000 Hunza inhabitants. We drove round the wide curve of an open and horizontal track, above the confluence of the Nagar and Hunza rivers, and across the watershed between the two states. Soon the track narrowed to a width less than that of the jeep body but just sufficient to accommodate the wheel base. Whilst the driver watched the passing edge of precipice on his right, I quickly withdrew my left hand from the doorway in fear of it being scraped by the jutting boulders of the vertical cliff towering above the nearside. I closed my eyes, feigning tiredness. Ahead, the track was being repaired at a point where falling boulders had gouged out a deep V-shaped gully. The army personnel implanting a new track section at the gap waved us quickly across, since a few stones were still descending from on high.

The wooden bridge over the Nagar river is suspended from wires stretched between stumpy pillars on each side of the gorge. The gap between road and bridge is not large, but I must confess that my moment of apprehension lasted more than the few seconds it took to cross onto the swaying and buckling bridge, which was only just wide enough to take the narrow jeep. However, it was not until we started to climb the steep zigzag track on the far bank that I became fully aware of the frailty of the track to Nagar. As the driver reversed round hairpin bends, close to the edge of vertiginous cliffs, I opted to walk part of the way in order to 'stretch my legs'!

The old kingdom of Nagar has a population of about 31,000, with approximately 6,500 people in Nagar itself. On arrival in the village we talked to the Tesseldar (chief administrator) and the visiting Assistant Commissioner about hiring porters, and later I visited the deposed Mir, a man of about sixty-five years of age, who was suffering from a heart defect. I was delighted to find that he was a man of wide experience and had an appreciative understanding of the conflicting policies in both national and international affairs. He explained that he listened religiously to the 6.00 pm BBC news because of its impartiality, and that his greatest wish was to see the peoples of his valley continue to prosper. However, in such a harsh environment as this, one could understand why his predecessors had banished miscreants, thieves and sometimes murderers up to the last village, Hispar, which seemingly served as some kind of penal colony. Decisions had always been the Mir's sole prerogative, although he relied heavily on the advice of his council of elders. This system of government had its faults, but few despots in any area of the Karakoram ever survived the first assassination attempt. The old feudal system thus had its own peculiar form of democracy.

Our conversation ranged over local diets, religion, tourism and literature, and the Mir's sporting love, polo – a robust, some would say deadly game in these hills, judging by the ferocity of play. He proudly told me that until Bhutto banned the game, his Nagar team were the local champions of Gilgit.

Later in the day, when everyone had assembled, David took the opportunity of holding a clinic. We had previously found the only occupant of the local dispensary was a very loud cockerel who had obviously now taken up a permanent residency. The smell confirmed our suspicions that he had been undisturbed for weeks.

The hiring of porters was a protracted business, but it was not without light relief. I feigned great surprise on learning that the journey to Hispar would take three days and not two, and that an equal number of days would be needed for porters to stagger back exhausted without loads after safely depositing our baggage.

The following day we left for Hispar at 5.30 am. It was a journey I shall never forget, since in 1957 I had listened to Shipton tell many stories about the area whilst we waited for a storm to pass us by. We soon entered the desolation of the deep ravine of the Hispar Gorge, but first we dropped down onto the Barpu Glacier, whose dirty moraines and thundering stream sweep right beneath the hanging fields of the village. Here the first party had witnessed a mud flash. The descent is really no more than a delicate line of footholes kicked into the loose cliff, and we climbed down onto the glacier as quickly as possible, trying to avoid touching the protruding boulders, which were held as if by some magical glue on to the decomposing cliffs of compacted grit that stretched for hundreds of metres above our heads. It was a relief to wander in and out and up and down the piles of moraine cones on the glacier. I felt as though I was home again. However, I knew that it was going to be a tough day. I had not yet acclimatised, and neither had I had any opportunity for serious training before the expedition. I resolved to get through the pain barrier as quickly as possible. Long experience had taught me that the only way to achieve acclimatisation and fitness in the Karakoram is to put a large pack on your back and then doggedly plod on, refusing to recognise any increase in discomfort or tiredness as the day wears on. Unfortunately Rana and Bob did not agree with my philosophy and hired porters to carry their baggage, and so for the next four days they unwittingly became the focus of my envy. Fortunately, David had similar views to mine, but this was little compensation since he strode on far ahead of me.

By the time I had climbed the far cliffs of the Barpu Valley, crossed the plateau and descended to the Hispar river and the only bridge before Hispar village, I was exhausted; but I knew now that after a quick breakfast I would be able to finish the day's walk. Hopefully tomorrow would be better. Matters were not improved, however, by a return of the squits and by the stifling heat in the ravine. We had already been warned that there would be no water-supply for two days, and so it was necessary to conserve our meagre supply.

Having crossed the bridge, we now travelled along the right bank of the Hispar river (looking downstream), taking care not to slip anywhere near the swirling shore line. The cool spray of the river, the beauty of the whirlpools, the crescendo of the flow as it struck against obstinate rocks, the white waters of the fast-moving and steep torrent, the few patches of quiet, brown water where it seemed the river was flowing back upstream and the unimaginable noise – all these were magnificent impressions, but they were soon forgotten when a boot slipped or a backpack lurched. Rivers such as these are merciless. Indeed, Paul and Tony were carrying a plaque to fix on a cairn on the Biafo Glacier to the memory of Pat Fearneough, who had recently been killed in the Braldu river in a zone similar to this.

Sooner than expected, indeed less than an hour from the bridge, we came to a flat, open piece of land containing a few boulders. Here there was no threat from the river and the cliffs above did not appear dangerous. This was to be our bivouac site for the night. I dropped my pack, and after a light lunch quickly prepared by David, I crept between two huge boulders that were cool to the touch and provided shade from the sun. There we rested until evening, at which time the shadows were leaping from boulder to boulder and the air temperature began to fall rapidly. While I was dozing, David had crawled inside his tent and had fallen asleep. Unfortunately, tent fabric does not provide insulation as boulders do – in fact, it intensifies the heat. When David awoke he crawled out terribly dehydrated, and was to suffer the consequences for the next two days. Supper that evening was a non-event, since no-one had thought to pack an optimus stove pricker or a spare set of spanners.

Musicians at a Hunza festival.

Porters cooking chapattis en route from Nagar to Hispar.

Eventually, after about an hour, I was able to flatten a pin and grind away its sides on surrounding stones to fashion a wire pricker for the hole of the fuel jet.

The next day we again got away to an early start. We saw no sign of the lone sheep we had seen on the far bank the previous evening, which had traversed down across scree slopes and must have fallen into the river. The scene was as yesterday; an unstable and chaotic landscape of brown, grey and light yellow, with an occasional sprig of thyme defiantly resisting the harshness of the desert landscape. We had little water with us, but some small respite could be gained by bathing one's face in the cold glacier waters of the river, if these were within safe reach.

The track was more or less at river-level, but frequently it was necessary to climb up and down 60 to 180*m* (200 to 600*ft*) high cliffs, to avoid spots where intervening ridges terminated in vertical rock walls plunging directly into the river. At one high point, the track crossed a scree slope over which we all ran to minimise the risk of being hit by falling stones. In the middle of the slope the track had collapsed and we had to climb down a crumbling wall and lower ourselves into a shallow cave. One slip here would have been fatal, but the bulge over our heads briefly shielded us from falling stones before we ran the final part of the gauntlet.

For breakfast we had no alternative but to use river water, since we were anxious to conserve the little clear water remaining in our flasks. As we continued I realised that although I was still trailing behind the main party, I was gaining in strength and could continue happily all day even if my pace was slow. David, however, was suffering from yesterday's dehydration and was now weak. Although he continued doggedly without a word of complaint, we had to call a halt in the late afternoon by a willow grove, which appeared as if by magic. Here the high grass, not unlike twitch weed, and the few branches provided a cool shelter, while the porters continued on to the camp close to a small beck. Within ninety minutes they were back with two full water-bottles and insisting that they be allowed to carry our packs. For once I could not refuse their generosity and did not feel as though I was thereby forfeiting my right to be travelling up these stupendous valleys.

We soon reached camp and crept under boulders, clutching cups of tomato soup prepared by Bob and Rana. After a cool bath in the long-awaited stream I felt the weariness of the past two days rapidly vanish and my appetite return. The third day started as usual, with the porters standing in a circle and singing prayers. The journey to Hispar was not long, and we enjoyed all the rest periods taken by our porters, who took the opportunity to smoke a weed which I was authoritatively told was locally-grown hash. I was just watching green and light-brown lizards with deep red throats playing amongst the boulders, when into our midst came a group of Polish climbers just returning from the ascent of two 7,000*m* (23,000*ft*) peaks. They looked a tough and happy bunch of mountaineers, and it was delightful to share food and mountaineering tales with them before they wandered off down the valley.

Soon afterwards we reached the bridge just below the village of Hispar, which complemented the three-day march in that it was safe and sound at both ends, but slightly hair-raising in between. Two wires provided the handrails, but most of the 5*cm* (2*in*)-wide, uneven slats of wood which the two base wires were supposed to hold were missing, and those that remained were broken or ill-attached. However, the sight of the river between the slats concentrated the mind wonderfully. There followed a long, steep zigzag track up to the beginning of the village, which was sited on the edge of a large alluvial fan, spreading out from the foot of a side valley gorge, like so many villages in the Karakoram.

A dramatic change occurred as we stepped over from the cliff path onto the fan. Suddenly there were flowers in profusion, high grasses and safe pathways twisting in and out to complement the wandering route of the irrigating water. Young boys crowded round, asking to carry my pack into the village for a few rupees, but now I was determined to persevere. The fragrance of the plants was almost intoxicating, and I could not help sitting down for a few moments close to a deep well from which emanated a strong, cool breeze. Eventually I walked on, followed by a crowd of inquisitive children, and crossed the gorge in which, far above, a glacier snout could be seen. Immediately around the next bend was the rest house. The first stage of a memorable journey was over.

The scene was not unlike other high and remote villages of the Karakoram. Wild roses of a slightly pink hue waved on sturdy bushes, as though to compensate for the huge and ugly snout of the Hispar Glacier only a short distance to the east. The menfolk wore flowers in their hats, and the children sat at their feet, occasionally handing us freshly-picked peas; the women, meanwhile, were safely hidden in their houses. David immediately set about treating the male population, also performing various pieces of minor surgery, and was later called to several of the houses of the village in order to examine some of the sick women.

It is very easy to become attached to these hill people and their way of life. They have no airs and graces; materialism is almost unknown among them; and because they are close to the great forces of nature, they are humble and friendly, wishing only to give the strangers in their midst a warm welcome. They were, however, rather upset that we had chosen Nagar men to act as porters up to our base camp on the Hispar Glacier, so I explained that we would employ Hispar porters on our way down.

That night we had a substantial meal, the first of the march, not only to give a backsheesh (tip) to our Nagar porters, but to celebrate the birthdays of two of my daughters. The only drawback to the idyllic setting was the horde of flies that infested every room of the rest house. That night I chose to sleep outside on a grassy bank.

We left at about 6.20 am the next day and soon were climbing the steep conglomerate cliff face close to the glacier snout. Here a villager returning from an outlying pastureland was operated on by David, who found a large foreign body embedded in the back of the old man's throat. Six minutes later we were in amongst the first moraine cones of the glacier.

Because they were close to the disintegrating snout, the moraines were unstable; the slightest misjudgement could cause a slip. Walkers in this terrain resemble dancers, skipping on and off unstable boulders, and side-stepping lumps of stone embedded in the ice as they make their way across the frozen surface. After an acrobatic hour we reached the far side and ascended a steep gully with the conglomerate cliffs towering over our heads. Fantastically-shaped, finger-like pillars of grit, stones and boulders looked as though they needed only a slight push to send them crashing down. Eventually we emerged onto a plateau of luxurious grazing land where we rested for a snack. Thereafter it was a series of similar but rising plateaux, one so large it could have accommodated a football pitch. On the last level we breakfasted amidst several crudely-built shepherds' huts, measuring approximately *2m* (*6ft*) wide and about *6m* (*20ft*) long, with the accommodation space being half below ground-level, presumably to give some insulation in winter months. The roofs were composed of crossed wooden rafters filled in with twigs, thatch and mud, and doorways were framed but small, being less than *1m* (*2 to 3ft*) high. Inside, the smell was almost intolerable, but the coolness and shade were sufficient compensation.

Immediately after our meal I wandered on ahead, thankful that my strength was returning. The track, which followed the ablation valley between moraine and valley wall, was now climbing again to keep parallel with the rising level of the glacier, out of sight to my right over the edge of the lateral moraine. These moraines bound each flank of the glacier which produces them as the ice-flow pushes forward. Occasionally the ablation valley narrowed, and the easy path wandered up rock tracks in the cliff face to rejoin the next section of the narrow valley above. Around one corner I entered a wide corrie (side valley) and knew that here was to be our camp for the night: ahead was a low, level, horizontal moraine at the foot of the corrie, that could only be the dam of the small lake we had been told to expect. Within seconds I was enjoying a delightful swim in shallow, brackish and not too cold water, and soon most of the team were splashing around too. It is strange how nudity and a common activity have the effect of welding a group together.

That night the mosquitos were troublesome, but before falling asleep I watched the numerous satellites crossing overhead. I counted four of these constant velocity objects in the space of half an hour, two on an apparent collision course. I wondered which of them were now communicating with our survey team.

The next day the ablation valley steepened as we approached the Khiang Glacier, a tributary of the Hispar. The vegetation also thickened and, judging from the size of old stumps, many substantial thick trees had once grown here. At the junction of the moraines, we were forced to climb a little way up the side valley, but once atop the Khiang moraine the descent down to the glacier was simple, as was the crossing. Midway over the glacier we repaired the stove again, had breakfast and filled water-bottles. After ascending the far moraine, which was steep and loose, we emerged onto flat grassland and, nervously edging around a wild but watchful bull, came to Bitanmal, a pasture area containing several shepherds' huts. The porters wished to stay here, but when we agreed to pay them for the next half-day they reluctantly consented to continue. In less than two hours we were in Dachigan camp, reunited with our friends, who gave us a welcome only those who travel the hills can appreciate.

After exchanging news and drinking several cups of tea, we surveyed the position.

George was obviously a very sick man; Ron had been bitten by a horse fly, and a sizeable portion of his leg bore the deep coloration and swelling of septicaemia; Professor Zhang was still on the glacier and immobile; the new radar equipment was not yet functioning correctly, and Gordon had to return home in a few days. However, the 1977-type equipment was performing satisfactorily. It was obvious that we had to move up to the ice camp two days' march away to continue our work. At this point, however, the porters became quarrelsome. Sad to recount, I have not yet read one story of Karakoram expeditions in which porters did not present difficulties. Fortunately, Major Rana could translate to them the seriousness of our position with regard to our sick comrades, and explain the necessity of getting all our stores up onto the ice so that we could conduct the experiments that Gordon, Lao Dong and Marcus were now convinced could only be done there. Arguments raged on all day, and despite an offer of a fifty per cent increase in wages, the porters returned to Nagar with a request for four Hispar men to come up and help us instead.

The Dachigan camp was a delightful base for the radar echo-sounding group. It was on a flat stretch of land with a little grass and sloping upwards in the direction of the Hispar Pass, wedged between the mountainside and the edge of the glacier. Here, after a slight rise, the moraine fell some 60m (200ft) down onto the surface of the glacier, while the hillside to the north had sufficient snows in its middle reaches to cause a little water to flow in its gullies. This entered our camp site in two streams about 20m (66ft) apart. One stream meandered close to the moraine, where it helped sustain a small area of woody willows, before finding a gap through which it descended onto and into the glacier ice. To complete the idyllic setting, flowers of many colours decorated the area immediately to the east of our camp.

The following day David, Paul and Bob moved up to the ice camp, whilst Tony, Marcus and I set out a survey base line and staked out several possible plane-table mapping locations. At one of these, high on a ridge above the camp, an eagle studied our every move, while we apprehensively watched its graceful flight and sudden changes of direction.

The next day, four men from Hispar arrived, and our doctor radioed from the ice camp to tell us to request a helicopter rescue, since the injury to Xiangsong was worse than a sprain and X-rays were required. This was a serious blow, since Xiangsong was not only leader of the Chinese team but also an outstanding glaciologist and had already started a number of studies that would have greatly enhanced the glaciological research work of the project. If a helicopter could be brought up, George, now a confirmed hepatitis case, would be flown out at the same time. If it could not, then it would take many days to get our casualties off the ice, down to Hispar village, and through the river gorge, since Xiangsong could not walk and the effort of standing for only a few minutes left George totally exhausted.

Until we received news concerning the availability of a helicopter, the porters were asked to rest for the day on half pay with all food and drink to be supplied by the expedition, to which they very reluctantly agreed. Major Rana asked if he, too, could return with the helicopter to Hunza, but this I refused on two grounds. We needed every available pair of hands to help us solve our problems, and he was also required to assist in translating the details of negotiations with our porters. Furthermore, to return to Hunza when he had only just arrived on the Hispar made no sense whatsoever.

Meanwhile, I had determined that plane-table mapping from our present position would be difficult on the scale required, since the base line was too small and far removed from the centre of our activities on the ice. We would have to rely on the theodolite readings taken painstakingly by Marcus during the previous three weeks while he helped conduct the ice depth-sounding experiments.

Soon it was confirmed that a helicopter was available, but the flight was delayed by twenty-four hours because of poor weather. On 7 August the helicopter came up the valley. It was almost overhead before we spotted it, dwarfed into insignificance against the backcloth of the huge Karakoram peaks. We all gave silent thanks to the pilots who had to manoeuvre their fragile craft in such difficult conditions; the air currents here are constantly changing due to fluctuations in air temperature, which in turn varies according to altitude, proximity to the ice and time of day. After landing at the ice camp first in order to collect Xiangsong, the helicopter then came downstream to Dachigan.

Whilst preparing a landing circle and talking to the helicopter by radio link, I noticed that Rana appeared to be sorting out some of his kit. However, I soon forgot this in the hurry of getting George ready, scribbling last-minute letters for home, talking to the ice camp by radio and settling yet another squabble with the porters.

When we saw the helicopter approach, we burned a little damp wood so that the trail of smoke would indicate wind strength and direction. It landed perfectly, bringing up a great deal of dust. George was helped aboard by Nigel, who had travelled up with the helicopter. As I waved farewell to Xiangsong, David Giles and Nigel, I saw Rana scuttle under the revolving blades and slide between the closing doors. The helicopter rose, and by the time the dust settled, it was close to Hispar village; another hour and it was over Hunza.

I was naturally sorry to have lost the company of four colleagues, but I was exceedingly annoyed at the desertion of Rana. While I do not doubt that he had to report on our innocuous research activities on his return, it was obvious that he did not relish repeating the rigours of the trail back to Hunza. I could appreciate that he had no desire to travel on the glacier towards the Hispar Pass and that his intermediary role was increasingly difficult to fulfil as he became more integrated into the spirit of our work; but I ceased to feel anything but resentment when I was told he had returned to Islamabad and Lahore to join the religious holiday festivals and attend a wedding.

We now decided to get as many stores as possible lifted up to Jutmau camp across the Pumarikish Glacier, so as to assist Paul and Tony in their attempt to reach the Hispar Pass. At the same time it was necessary to retrieve equipment from the now-deserted camp on the glacier opposite Jutmau. Gordon now had two days left in which to modify his electronics equipment on the new radar set so that it could function correctly on Karakoram ice. Marcus and I had to try to sort out the survey data, whilst Ron had to rest for at least a couple of days and take the weight off his still swollen leg. Around noon the Hispar porters left for the higher camp, complaining that their packs were too heavy; to rebut their arguments, Tony picked up each pack one by one, using only his right hand. Lao Dong and Bob, stranded at the ice camp when the helicopter left, had now started their way down the glacier, threading a path through endless stone and boulder cones of high moraine that completely covered the glacier surface. After the helicopter rescue, there was no food left on the glacier and it was important that they reached Dachigan that evening.

Meanwhile, Paul rested after the exertion of the past few days. He, Tony and Ron had performed Herculean tasks during the previous three weeks, transporting heavy generators, fuel, radar instruments, food and tents to various locations on the glacier where the depth of the ice was being measured. They had also had to reconnoitre through the moraines, over the ice and across melt-streams and crevasses, and often then to seek safer alternative routes – a never-ending task.

As the day progressed, our anxiety for Lao Dong and Bob increased. They had only a little Alpen to sustain them, and Bob's voice on the radio was beginning to indicate more than weariness. Because of his experience on Everest I initially gave the matter little thought, but as time passed and his voice echoed more and more concern, the tension increased. I had already noted that Bob had been in poor health on the way up, and was probably not fit. Furthermore, he was not sure of the way, and indeed by this time he knew there *was* no identifiable route and that it was simply a matter of climbing one moraine cone after another and hoping to make steady progress downstream through the maze of rock piles which stretched all the way down from opposite Jutmau, past our camp to the very snout of the glacier. It was with great relief that they eventually recognised their position and wearily climbed up off the glacier into the ablation valley where our camp was located.

That night it rained, and later the next day the porters arrived back from Jutmau, having done nothing to help retrieve the kit from the ice camp. Their complaint was that the journey was three days, not one, and hence they deserved a total of six days' pay. I told them that Paul had previously done the journey there and back in one day, but it was to no avail. In addition, they wanted two days' full pay for walking up without loads from Hispar, a journey that had obviously taken them only one day, and they also demanded similar rates for the return journey! Under no circumstances would they help us over the crucial next four days, and so, after twelve hours of discussions, I had no alternative but to send them home without pay, informing them that I would report their unreasonable behaviour to the Tesseldar in Nagar.

Meanwhile, Bob had descended to Hispar with Gordon in eight hours and had sent up a porter to collect his kit while he took photographs in the village. Because of deteriorating relations between ourselves and the Hisparis, based no doubt on the biased views of the returning porters, Bob was, however, forced to leave the village in some haste and was relieved to reach Nagar. Gordon, on the other hand, had no reason to tarry in Hispar and after only a few hours' sleep he had quickly moved on down the valley.

In spite of everything, moving heavy research kit down and across the glacier was welcome exercise prior to our next series of experiments. That night we fully expected the Hisparis to invade our camp site, so black did the situation appear from Bob's hurriedly written note. Fortunately nothing happened, and so Paul and Ron left for Jutmau to relieve Tony, who had been there on his own for two days. We still had to remove the kit from the ice camp, and in the poor weather we were now experiencing, we knew this could take at least four days.

By the next morning I had concluded that we should withdraw the glaciological unit to Hunza and concentrate our remaining efforts on the Ghulkin Glacier, where lines of communication were shorter and less hazardous. Gordon had by this time started for the UK, and Marcus was poorly, probably sickening with hepatitis. There was no way I could rely on the local porters, and so on the next radio call I asked Nigel to send up a group of Nagar men plus Major Rana, if available, or a policeman to help us sort out our troubles in Hispar. Marcus, Lao Dong and I then set about packing our kit and dumping hidden supplies of food for Tony and Paul, who still wanted to make a dash for the Hispar Pass and a descent of the Biafo Glacier. Similar dumps were prepared for a unit of the survey group who wished to come on a holiday visit to the glacier at the termination of their work, time and weather permitting.

For the last time, we moved into the glacier moraines opposite Dachigan and retrieved all the electronic equipment there. One final series of experiments – made possible only by superhuman efforts by Gordon in his last forty-eight hours at Dachigan – was conducted, using a composite electrical unit from both the new and old systems. Unfortunately this proved fruitless, presumably because the signal was too weak; nevertheless, during the overall period spent on the glacier, we had made sufficient ice-depth soundings with the updated Vatnajökull equipment and had sufficient theodolite survey results to allow us to compute the locations and determine accurate ice-depth profiles of a major Karakoram glacier for the very first time. Initial results tentatively indicated that the ice-flow could have a maximum depth of about 600*m* (2,000*ft*), and that the glacier had gouged out a deep, long, but horizontal valley. This information, plus the discovery that our signals would pierce the thick moraine cover on top of the ice, was of some academic importance; if we could do similar work but of a more practical significance on the Ghulkin Glacier, then the ice-depth project would be totally justified, despite its depleted membership.

The last two days passed quickly – none of us knew what day of the week it was. Ron came down, carrying a huge load from Jutmau on his own – a risky venture, since one particular glacier moraine stream was very dangerous to cross and was probably the cause of all our problems with the porters.

When the recovery party of twenty-five porters arrived, we gave them loads and despatched them down to Bitanmal, where they wished to sleep the night. At the same time, as though to restore my faith in Hispari men, two local hunters came across the glacier and presented us with a goat's leg. They would take no payment, but gladly accepted various items of kitchenware, tea, spices and empty cans. We then selected three of the twenty-five men to go to Jutmau the next day to bring down the kit which Ron had left neatly stacked on the side of the glacier. Marcus was to descend with the Sirdar, the leader of the porters, and select eight more porters to carry down our remaining excess kit.

Two days later, Ron, Lao Dong and I started on the descent to Hispar. Despite the fact that he was still quite ill, Ron took every opportunity to film, and even recorded the singing of the porters. At that time we were unaware that Ron, Paul and Tony were all three already infected with hepatitis, and that their first symptoms were beginning to show. In deteriorating weather, Paul and Tony were now making their attempt on the Hispar Pass.

We arrived in Hispar late in the afternoon, and immediately the haggling began. I informed the headman that he should never have approved two of the renegade team of four Hispari porters, since one was too old and had admitted he could not carry even a 15kg (33lb) load, while the other was obviously seriously ill and needed treatment for a hernia. Meanwhile, unknown to me, the headman, trying to be fair to all his friends who were hoping to be hired and collect high wages, had selected another group of porters to replace those who had now helped us down from Dachigan. The chaotic scene can be imagined. The argument raged all night, long after I had clearly stated that I would make my decision in the morning and had retired to my sleeping bag to feign sleep, while trying to devise some equitable solution.

By the time we had consumed a wholesome breakfast of eggs, Mars bars and tea, the best compromise solution had emerged and one on which the headman, the policeman, the Liaison Officer of another expedition and myself all thankfully shook hands – studiously ignoring the gloomy faces of the men who had lost out in that compromise. Needlesss to report, the new team did not include the old renegades who had presented themselves as ready and able for duty.

The bickering continued during packing, however, and so eventually I despatched our team on their way and remained behind to try and soothe those who still felt aggrieved. By the time I reached the steep track down to the bridge, I had caught up with everyone except one old porter, who had charged ahead of the rest of the party and had descended the cliff track and crossed the bridge to the right bank of the river. He refused to hear our calls – wisely, as events proved – and continued down the track we had used on the way up, even though we had been told this track had been temporarily destroyed. It appeared that he did not approve of the decision by our porters to travel down the left bank of the gorge, which I was told was a faster route to Nagar.

The track slowly descended to river-level, and soon it began to rain, very slightly. The porters became a little agitated. At one point they stopped for a 'five-minute rest', but refused to move on for another thirty-five minutes. At the next stop, under a huge boulder, where small caves had been excavated and stone walls built to provide several individual shelters, another 'five-minute rest' was requested. After ten minutes I announced that I would carry on rather than wait; but several kilometres later, I could still see no sign of movement from behind, and as I was unsure of the route ahead and unable to find any shelter from the steady drizzle, I decided to go back and find out the reason for the delay. At the edge of a large scree slope, wet and slippery and too close to the river for comfort, I was cautiously jumping from boulder to boulder when I momentarily stopped and looked up. There, descending upon me, were several stones; one I managed to catch and drop, one I deflected with my hand as I ran – but now my eyes were focused on a falling boulder the size of two large buses, still far above. With no time to jettison my pack, I leapt from stone to stone, all the time trying to keep one eye on the rapidly advancing black mass. As it bounced, creating a hail of smaller, but still large stones, it changed direction, forcing me to turn and retreat in hops and jumps of giant proportions and great rapidity. Events moved so fast now, and my reactions so slow by comparison, that I did not have time to notice anything other than the fact that the huge, black rock passed me whilst I was still running towards it. It crashed into the river without increasing the noise-level of the torrent, and instantly disappeared without trace.

I turned again and beat a hasty retreat to the nearest edge of the scree, dodging an avalanche of smaller stones, and fell exhausted to the ground. Several score of smaller stones had passed me, some of which I had seen, but many of which I had simply heard, whistling past my ears. My lasting memory, however, was of the huge boulder as it passed me, momentarily poised in space less than 9m (30ft) away and completely blocking the view downstream. This view, to which I had become accustomed, was one of brooding inhospitality as the swirling mass of water surged steeply down, to disappear into the lower sections of the gorge. My feeling for the river now changed to undisguised horror; I saw it as a yawning trench in which nothing lived and from which it would be almost impossible to escape. Now I knew why the old porter had crossed to the right side of the river, even though he had been told that the track had been obliterated, and also the reason for the rests and discussions by our left-bank porters as soon as the rain had started. Obviously they had been debating whether or not to turn round.

Within half an hour I had rejoined our group to cast my vote and found that they had not moved from the 'five-minute rest' position. To say I was angry at not being informed of the high risk of stonefalls on the left bank is an understatement; but I was doubly annoyed that they had permitted me to journey on alone. In the event, we decided to carry on, since we had come too far to turn back now, and we had already crossed several zones that would be just as treacherous as those that apparently lay ahead. The few drops of rain now falling were the cause of the damage. In effect, the raindrops were just as deadly as an earth tremor, since they could remove the last few grains of dust holding up several of the millions of tottering blocks far above, and so create a series of boulder avalanches.

Several porters did not agree with the majority decision, but after lengthy prayers they shouldered their packs and, with grim faces, headed downstream. The journey now followed the river-level, and frequently we had to edge along loose blocks jutting into the ever-turbulent stream. I now could not travel more than a few metres without glancing up at the towering cliffs, and this nervousness was to remain with me for the rest of the expedition. Fortunately, however, the porters knew all the danger zones. On reaching such places we all stopped, the porters would say prayers, examine the slopes above, listen for whistling stones, and then, one by one, when it was judged to be clear and the person ahead had reached a safe point, we ran as fast as humanly possible across the dangerous stretch. Immediately after one long run over a loose sandy scree, a blitz of several score of stones screamed across our tracks. It was exhausting work, and the day was long.

Ten hours and several frantic runs after leaving Hispar, we approached the base of a wide gully that opened out higher up into a wide plateau; there, perched incredibly 500m (1,600ft) above the river, were cultivated fields. The steep ascent took a long time; no-one reached the outer walls without several pauses, but to have sat down to rest would have drained our last reserves and our determination to reach the haven, now so tantalisingly close. At last we felt safe and secure. The fields of Huru were maintained by one old man, Hassan Ali, who lived there on his own all the year round. He sold us nuts, milk and apricots and joined in our celebrations at having descended more than half of the gorge from Hispar to Nagar. We were informed that the remainder of the route was free of danger.

The contrast of Huru was most welcome. Gone from our ears was the constant noise of the river; we could talk softly now and be heard. Two small, gentle streams of water, one clean and one heavily dosed in iodine, issued from springs only a few metres above, the latter possibly being the cause of the old man's goitre. In the centre of the village, a pond had been excavated and here a large willow tree gave shelter. All of us were exhausted, and so we sat and munched a little food before erecting tents and brewing up.

The following morning the track climbed higher and ever higher. On the other side of the gorge, stream-like layers of cloud clung close to the valley walls, drifting upstream, many of them below our present position. All was quiet and still, but the sense of an evil presence still rose from the hidden depths as we climbed higher and higher, as if to escape yesterday's demons.

Later that day, Lao Dong, Ron and I crossed the Barpu Glacier, climbed the fragile cliff and entered Nagar, where we learned that Bob, Gordon and Marcus had arrived safely several days previously. With that good news we took a jeep and rode in style to Aliabad to meet our friends.

Meanwhile, Tony and Paul were grinding to a halt just below the Hispar Glacier Pass to the Biafo Glacier. For a party of only two mountaineers to make such an attempt was a little risky, but both were veterans of Karakoram expeditions. Tony had filmed high on K2, and Paul was soon due to attempt a winter ascent of Everest. When they left us both appeared to be very fit and fully acclimatised. They knew the risks they were taking and I felt confident that they would not put themselves or the expedition in jeopardy. Their main problem was pack weight, each having to carry 34kg (75lb), consisting of ten days' food, a tent, climbing rope, film gear, cooking and sleeping equipment – and this at an altitude of 4,300m (14,000ft). But the compensation was views of peaks such as Kanjut and even the Ogre peeping over the Pass. Tony writes:

'Beyond the Jutmau Glacier, the track petered out again into bad scree and we descended to the Hispar Glacier to make our way out to the middle, hoping to find bare ice to walk on. I was already savouring the isolation and wild Karakoram mountain landscape. No gentle Himalayan snow domes here – all the classic giants are complex structures, either with phenomenal red rock buttresses and towers, or enormous convoluted snow and ice faces, where extremes of heat, cold and gravity combined to freeze hanging masses of ice into impossible-looking forms. We slept in a moraine depression that night with light snow falling.

'It was easier when we made the bare glacier ice. I can't remember talking to Paul at all that second day – one of the advantages of knowing each other well. We pitched the tent and made a small dump of food and film on a rise which gave us a clear view of the glacier up to the Pass.

'The luxury of a firm surface to walk on lasted nearly to the bottom of the Pass. After a civilised cup of tea we put on a rope and set off to try to cross it. Very soon Paul slowed down and began moving from side to side, probing the snow and trying different lines to avoid crevasses. It was hopeless. The whole area was honeycombed with cracks and very dangerous, the soft snow cover being just deep enough to make crevasses invisible. It also began to snow, so we descended to the level glacier and camped for the night.

'The next two days (or was it three? or four?) became a collage of mixed memories. A foot or more of snow fell, and time became distorted; eating, sleeping, talking; occasionally stretching our legs. I delved into our one luxury, Thomas Hardy's *The Return of the Native*. Between us we sorted out life, literature, literary criticism, the criticism of criticism and the value of value judgements. I can't actually remember what we decided, but we seemed to be in agreement. I think we probably sorted out history as well; we usually did.

'The snow increased the dangerous condition of the Pass, and the inevitable decision had to be taken to return, so we began plodding back. The soft snow had totally changed the ground conditions, and we remained roped together and alternated the lead, probing every step with a ski pole to locate possible crevasses. Visibility varied considerably but we eventually made it back to the gear dump and then camped at the edge of the moraine with superb views of Kanjut. Next day we stayed on the glacier to skirt round the junction with the Jutmau, and exited, exhausted, at the site where we had left Ron.

'We thought the adventure was over, but the sting in the tail was to come on the tracks through the ablation valleys to Dachigan. Once there we knew that a few hours further would take us to the base camp from which Dave Wilkinson was leading an expedition to climb Khinyang Chhish via a new route. Another glacier crossing took us to his deserted camp. I ferried equipment from Dachigan, while Paul made a solo trip to try to find their route. He arrived back that evening with a damaged hand, having survived a slip on loose ground. We were both feeling more and more jaded. Suddenly Dave's team appeared for a rest and more supplies, having made good progress on the new route. We felt too listless and short of time to join them. Unknown to us, we had contracted hepatitis, and it turned out later that they caught it as well.

'After they had set off back up the mountain, Paul and I became feverish. Luckily, our bouts alternated, so that one could heat drinks while the other was semi-conscious. We knew we had to get out fast, so we set off on the nightmare walk out. The baking hot valley became worse as we lost altitude, and three days later, thirsty and weak, we arrived at Nagar and were able to telephone the project base camp at Aliabad. They must have set off almost immediately, because we were still enjoying the novelty of trees, shade and good water when a Land Rover arrived and we were overwhelmed by a warm welcome from John Hunt, Ron Charlesworth and Bob Holmes. Going back in the Land Rover I leaned back and closed my eyes.'

# Journey to Shimshal
by Lord Hunt

'I'm intrigued by your reference to "residual ambitions",' wrote Keith from base camp, in reply to my letter about the arrival of my wife Joy and myself to join the expedition during its last few weeks. 'A summit, perhaps? or a journey to Shimshal?' I confess that Shimshal was not one of those 'residual ambitions' which I had entertained before returning to the Karakoram after an interval of nearly half a century. A peak, however modest, was likely to prove more tempting, with its prospect of viewing once more those high summits in Baltistan where I had climbed in 1935. I had, in fact, forgotten that it was in Shimshal that Eric Shipton and Michael Spender, in rags and tatters, were received by the astonished emissaries of the Mir of Hunza in 1937, after descending from the pass of that name at the end of an arduous journey in which they had surveyed the vast area on the northern fringe of the Karakoram, described by Shipton as a 'Blank on the Map'. The dignitaries of the local ruler had not expected to act as escorts to these odorous vagabonds.

To those who have travelled in the mountainous region where the frontiers of four nations run, the word 'Shimshal' (also spelt as Shingshal and Shimshall) conjures up a 'Shangri-La', a haven isolated by distance and natural obstacles from the nearest habitations in the Hunza Valley, and from the dwellers on the high steppes of the Pamirs in Chinese territory. It is a settlement where human beings can, even today, pursue a peaceful, pastoral existence untouched by the forces of change. It is, however, not easy to reconcile the gentle farming folk whom we were to meet with their brigand forebears who, less than a century ago, were still following 'the gentlemanly profession of caravan-raiding and man-stealing',[1] in the tradition of their kinsmen in the upper Hunza Valley, by waylaying travellers carrying merchandise between Yarkand and Ladakh.

Apart from Shipton and Spender, very few westerners have visited the village since the British established their authority over the rulers of Hunza and Nagar in 1891 in pursuit of a 'Forward Policy' to check Russian designs on India and Afghanistan, and thus to settle the Central Asian question at that time. Sir George Cockerill made his way up the gorge in 1892; in 1925, a Dutch pair, Dr and Mrs Visser-Hooft, whom I had the pleasure of meeting later in Srinagar, visited the village; during the 1930s Mr and Mrs Lorimer also paid a visit during a prolonged stay in the village of Aliabad; Schombergh was there in 1934. Since the last war we were told of only two more visits by Europeans, both of them in 1974. The prospect of becoming the latest in this small and select company was too tempting to resist, and we quickly accepted Nigel Winser's offer to make the necessary arrangements.

But I was also keen to realise that other ambition: to reach a viewpoint which might open a door to distant memories. We first spent two days with Keith Miller and Marcus Francis in their camp beside the Ghulkin Glacier, where they were

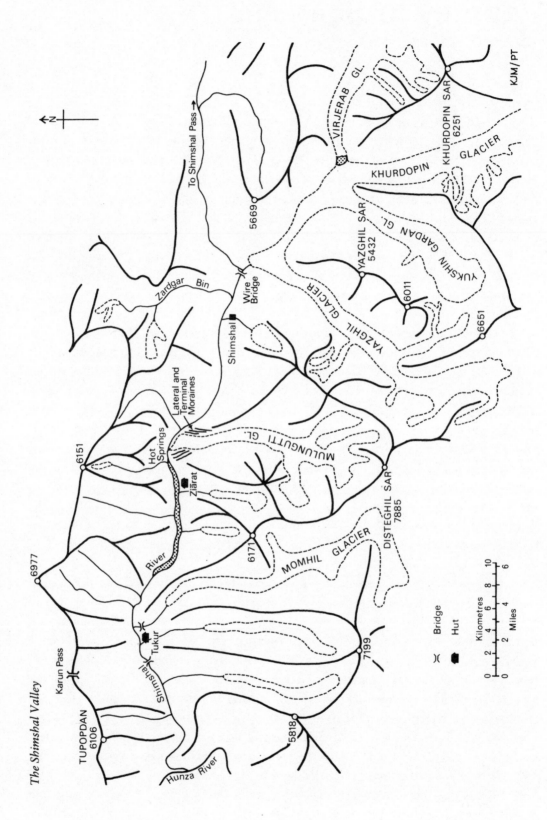

surveying and measuring the glacier's depth and rate of movement. The little oasis of tamarisk, juniper bushes and wild roses in its setting of soaring peaks at the head of the valley was just what was needed to make us feel a part of the expedition and the superb environment in which its members were pursuing their studies.

Back in the village of Ghulkin we spent a delightful hour with one Gohar Shah, who had recently returned to his home from the British expedition to K2. Over tea and cakes we told the assembled company of villagers of our plans to visit Shimshal. There was great enthusiasm to join us; not only, I sensed, as porters, but as our companions. 'You must take men from Ghulkin and Gulmit', they said, 'The men of Pasu are no good.' Indeed, we needed no persuasion, for we had just spent two days in the company of some of these people and were able to appreciate their qualities. 'How long would the journey take?' we asked. They looked with doubting eyes at Joy and me and expatiated on the difficulties of the track. 'It can be done in three days,' they replied, 'but for you, it will take four, even five days.' It is salutary, if a trifle sad, to be reminded that the spirit may still be willing, but the flesh is perceived as weak.

Two days later, however, I was able to restore my morale by making a rapid ascent of Hachindar, which rises 2,100m (7,000ft) above the base camp, at the great bend of the Hunza river, opposite Rakaposhi. Its height is only 4,570m (15,000ft), a modest peak indeed, and one which denied me that hoped-for view of the great peaks as I arrived there in drizzling mist. But it was also a kind of pilgrimage, for this was one of the survey stations established by Kenneth Mason nearly seventy years before. I had stayed with him at Oxford after returning from an expedition to Saltoro Kangri in 1935, in which he had shown a great interest. It was also an opportunity for me to appreciate the enterprise and skill of the surveyors with this expedition, who had identified and replaced a number of Mason's cairns. I felt, too, that marvellous sense of achievement which everyone must experience on a summit, be it Rakaposhi or Snowdon, as I caught a fleeting glimpse of the base camp far beneath me, beside the glinting river, through the swirling clouds.

On 2 September our party left base camp, escorted by the indefatigable Nigel, and well provided for by Shane's efficiency and thoughtfulness. We were on our way to the village of Pasu, to begin a round trip for which, despite the pessimistic prognostications of the men of Ghulkin, we had allowed only seven days. It was in the character of this international project that we comprised a cosmopolitan group. Apart from Joy and myself there were Chen Jianming, a Chinese surveyor and geologist; Ashraf Aman, a Pakistani mountaineer who had achieved fame by climbing K2 with a Japanese expedition in 1979; and Dr Robert Muir Wood, a Cambridge geologist, who had joined our expedition as a journalist to cover its activities for the *New Scientist*. They were delightful fellow-travellers. Robert's reading of the rocks added greatly to our interest; we much appreciated his acceptance of two veterans as his friends, and his six feet four inches, topped by a straw sombrero, were sometimes a useful marker for laggards. Ashraf's constant care for our creature comforts and conversational skills won him two more friends to add to the many he already had. As for Jianming, although our ability to exchange information and ideas was minimal – which, in view of his previous knowledge of Hunza, I greatly regretted – I have seldom travelled with an easier and more companionable person. He walked all day, self-possessed, a familiar figure in blue uniform and cap, a preposterous rucksack on his back, taking an evident interest in much that was beyond my knowledge. His manners at meal-times put us hungry Westerners to shame.

At Gulmit we stopped to pick up four porters, and, I suspect as a concession to the elderly element in the party, a cook. They were an interesting and contrasting company, some wearing clothing which betokened previous experience with expeditions. Serajuddin, tall, slim and wiry, with a twinkle in his grey eyes, sported a TOG 24 jacket, the very latest fashion among British climbers; Haidar Ali, a contrastingly short, stocky, bald and solid character, wore boots and breeches which proclaimed a French connection, as did those of our cook Sher Khan. Shaban and Mahomet Rajab were of a retiring nature; Rajab was even a trifle 'fey'. He it was who, a few days earlier, had performed before our gaze a frenzied dance during an entertainment at Aliabad. On that occasion, his face daubed with the blood of a slaughtered goat, he had whirled in a kind of escalating ecstacy, pausing now and then to receive messages from the drum and pipes of the enthusastic village orchestra until, brushing through the thicket of encircling crowd, he collapsed in a kind of coma. His performance appears to have been remarkably similar to the experience of Schombergh when, forty-six years before, he referred to feasts of Hindu origin that take place when crops are sown and at the harvest. These included goat sacrifices and dancing to music, with the 'yuderin', or fairy drum, among the instruments. Schombergh watched a female soothsayer – the 'Dainyal': 'She halted by the band . . . and listened first to the pipe and then to the drum, for a fairy voice, to be heard only by her' and seemed to become genuinely 'possessed'. Fairies figure a good deal in local folklore, and it is worth recording that a century ago the Mirs of Hunza were believed to be descended from a union between Alexander the Great and a fairy. As we looked at this quiet, retiring man, shouldering his load before beginning our journey across the half-finished suspension bridge over the Hunza river above Pasu, it was difficult to believe that he could be the same person whom we had seen dancing.

For the remainder of that first day we were travelling up the forbidding canyon through which the Shimshal river has carved its way to join the wide reaches of the upper Hunza Valley. It was a journey memorable for overpowering sensations: there was the uproar of the river, echoed and enhanced by the narrow confines of the towering cliffs, which in places almost shut out the daylight; there were the hazards of the track itself, improved though it has been with money provided by the Aga Khan, which calls for constantly careful footwork and a steady head in places. At times we were led down to the boulder-strewn stretches of sandy beach beside the water; at others, we had to climb for several hundred metres, to traverse vertical 'paris' or rock walls, following narrow galleries blasted into the cliff face. It was a breathtaking, awe-inspiring experience. That feeling of awe is my enduring impression of that long day, and it was a feeling increased by the dismal weather: clouds shut out the limited view and from time to time we were damped by drizzling rain.

Avalanche on the Batura Glacier.

A wasting ice-cliff on the Ghulkin Glacier.

*Left* Crossing a melt-stream on the Hispar Glacier via a giant boulder; *right* David Collins sampling the Batura Glacier melt-water; *below* Tony Riley en route for the Hispar Pass, weather deteriorating.

Marcus Francis (*right*) with the radar echo equipment on the Ghulkin Glacier and (*below*) examining a film record of an echo.

After several hours we stopped for lunch on a beach at the entrance to a shallow cave in the marble cliffs. This gave us our first chance to appraise the qualities of Sher Khan, our cook; a smallish, lithe and lively person who, after quickly restoring us with hot soup, proceeded to demonstrate his agility as a cragsman, almost running up a fairly hard rock pitch to inspect a small cave.

We had heard with some foreboding of the need to ford the rushing torrent at one place. It was, therefore, a relief to discover that at that time of year it amounted to no more than wading knee-deep, close to the rock wall, in order to cross a break in the cliffs. Reluctantly, but at the insistence of our men, Joy and I allowed ourselves to be carried at this point. Immediately beyond, a *6m* (*20ft*) high slab, steep, smooth and wet from the splashing of the river, provided the only obstacle which called for any climbing skill; a wire rope was a welcome hand-hold here. We had also heard that, at another point, we should have to cross the river *à la Tiroléenne* on a steel hawser; in the event, we found a remarkable bridge, of truly Heath Robinson design, its footway suspended from steel ropes and paved with loosely placed slabs of rock and pieces of timber. It is near this place that an alternative and much more arduous track descends from over the mountains on the right side of the gorge. This involves the traveller in a climb of 1,830*m* (6,000*ft*) to cross the Karun Pia pass and is the only feasible route when the river is in spate.

Seven hours after crossing the Hunza bridge near Pasu, we reached the confluence with the Momhill torrent. Here, two simple buildings provide a rude overnight shelter for travellers. The place is known as 'Tukur', meaning a bridge or a boundary in the Wakhi language. A little later, we were stretched out in our sleeping bags around an open fire in the centre of one of these huts, while Sher Khan cooked our evening meal. It had been difficult to converse throughout the day owing to the din of the water and the exertions of the track; now was the time to relax and recall the Wagnerian drama of the Shimshal Gorge.

The landscape through which we continued our journey on 3 September was no less spectacular, but more varied and less oppressive. The valley begins to open out, the limestone gives place to slates, the water is less deafening. We caught glimpses of high peaks up side valleys, for the clouds had cleared to give us blue sky and sunlight, which dispelled the gloom of the previous day. We traversed a low col dividing two tributary streams and descended to cross one of them by another remarkable bridge which spanned a deep canyon; it was short but sensational, built of springy boards and without any handrail to help us steady ourselves. Then followed a steep climb to traverse more cliffs. Further on again, we began a series of traverses across immense screes of steep, loose shale which started their journey thousands of feet above us and plunged into the river far below, to be undercut and washed away. I have never seen such screes as these; they bear striking testimony to the instability of the Karakoram landscape. Some of them must measure about 1,800*m* (6,000*ft*) vertically from the lofty crags above to the river-level. For some travellers at least, the initial experience of crossing them can be mildly unnerving; the more so because the moving surface quickly removes all trace of the path.

In fact, the journey on that second day remained trackless over long stretches, for after plunging down from the last scree, we found ourselves boulder-hopping over the last 5km (3 miles) close to the river's edge, in a wide open plain. It was pleasant to talk and be heard; I enjoyed Ashraf's company and was able to appreciate his knowledge, based on his own travels and reading, about his people and their history. He told me of the legend about a giant who once terrorised Shimshal and fed upon its inhabitants. The story ran that, when he had eaten them, he descended to Pasu and the other high villages in Hunza and helped himself to meals at their expense. There are several tales of man-eating giants in these parts, related by Professor Jettmar and others; this one sounded to me like a 'scare' story used by mothers to deter their errant offspring. Ashraf is a famous mountaineer who is employed as a local guide by the Pakistan Tourist Board, and we spent much time discussing the development of tourism and mountaineering in Pakistan, about which he held strong and doubtless – as is the way with climbers – controversial opinions. He gave me a foretaste of the history of Shimshal, about which we were to learn more after our arrival.

Our resting place that night was a caravanserai – three well-built, single-storey stone cabins, on a terrace above the dry river bed. They were in the traditional style of Hunza architecture, with sleeping platforms around a central hearth, on which rugs were spread. It is remarkable, as Ashraf pointed out, that these comforts, as well as cooking pots, remain in the caravanserai for the benefit of the traveller, and the doors are always open. A Shimshal villager, Rahimulla, who had accompanied us all the way from the Hunza bridge, had gone on ahead and with typical hospitality had tea awaiting our arrival. After supper that evening, while the men baked chapattis on the open fire and Sher Khan busied himself over the primus stove, Rahimulla sang us a Shimshali song. Translated from his native Wakhi language, we were told that it was about a man's love for his home, which, according to the Shimshali bard, he valued more highly than the palace of the Maharajah of Kashmir; in our own idiom, 'an Englishman's home is his castle'.

This serai is known as 'Ziarat', which means a shrine. According to Schombergh, it commemorates a Muslim saint, Shams-i-Tabriz, who lived in the time of the Emperor Sháh Jehan (1592–1666), for whom a miracle was performed after he had been shunned by all with whom he came into contact, on account of his evil smell. This account scarcely explains the annual pilgrimage of large numbers of Shimshalis, women as well as men, to make their votive offerings here in the month of October. These appear to be intended for some other deity, half-forgotten in the mists of a time long ago, before the advent of Mohammedanism. Perhaps this was the female goddess, Murkum. We were told that the purpose of the pilgrimage was to ensure fertility and health for mothers and children and successful crops for the farmers, which Professor Jettmar believes to have been the particular gift of this goddess. She was also said to have been an essential intermediary for the hunters of ibex and markhor, which are still fairly abundant in these mountains. Murkum regarded them as her flocks and herds. If a man wanted to shoot an ibex, he must refrain from sleeping with his wife the night before; whereupon the goddess would tell him in a dream where the herd might be found. The presence at this shrine of a number of ibex horns does lend credence to this superstition.

The upper part of the valley through which we continued our journey on the third and longest day provided yet further contrasts of scenery. The first few kilometres of rather dull plodding along the wide, featureless river plain led to an area of vegetation, so rare in this desert land save where man has by patient labour irrigated the soil as to seem almost a Garden of Eden. We were climbing up through the ancient moraines of the Mulungutti Glacier, watered by a clear stream beside which grew a small forest of junipers and birches. Prominent among the undergrowth in this paradise was a bush bearing bright yellow berries; roses, too, abounded, though we were just too late for the best of their flowering. Birds and butterflies, of which we had seen little since our arrival, were plentiful. The only thing lacking was sunshine to light it up. After a bright start to the day, clouds had now descended over the summits. A little later, after crossing more recent moraine ridges, and as we at last began to cross the glacier, described by Schombergh as that 'fine sheet of ice with Dastoghil at its head', the magnificent view of that 7,885m (25,868ft) mountain, now known as Disteghil Sar, was denied to us.

The climax of that day was our first sight of Shimshal. We had crossed the many ice ridges of the glacier and ascended a steep moraine slope, up which the track had been carved through a gully. At its top, the whole final stretch of the valley lay before us and several kilometres distant, an emerald patch lit by a shaft of sunlight filtering through the clouds, was the promised land: an oasis of plenty set in a vast, empty, ochre-coloured desert. Despite the distance, Shimshal beckoned us on and we were still relatively fresh when we arrived, after nearly ten hours' march. On the way we witnessed a remarkable spectacle. On the far side of the valley a huge slice of mountainside, forming part of an alluvial cliff and perhaps as much as 300m (1,000ft) in height, suddenly collapsed. With a deafening noise which drowned the roar of the river, a pale-brown avalanche comprising hundreds of tons of rocks and earth was disgorged into a natural gully in the cliffs, thundering down to the level plain and sending up a dense cloud of dust almost as high as the parent cliff itself. It was a sobering reminder of the fragility of this precipitous country and of the hazards to which its inhabitants are exposed.

Our arrival in the village was unspectacular; there were few people around. Two local villagers had been travelling just ahead of us and may well have reported that we were on our way, but although we were certainly the first white people to come to Shimshal for at least six years, we were taken for granted; it was, of course, as we would have wished it. An elderly person came across the fields and soon three or four others joined him. After friendly greetings we were led to a small painted bungalow, different from the generality of Shimshal houses. It had a verandah, in front of which was a vegetable garden, and within its two rooms were beds, tables and chairs and windows boasting panes of glass: a luxury in these parts. Within an hour other villagers arrived. They included the Lambardar (headman), Gol Badan, the schoolmaster Daulet Amin and his 'star' pupil Johar Ali, on vacation from his studies at Karachi University and the only local person to achieve that distinction.

Another arrival, just as the light was failing, was a small caravan of yaks carrying ghee and wool, upon which some children were riding. They had come from the high pastures near the Shimshal Pass, where herds of animals are kept throughout the year. For Joy and myself, they provided a link with Khumbu, some 2,400km (1,500 miles) distant in the eastern Himalaya.

Next morning, accompanied by the Lambardar and our other friends, we made a tour of the village. There are, in fact, three colonies, each on separate terraces and irrigated from different glacier torrents. Tukur, which we visited first, stands above and beyond the main village; Shullallah is downriver from Shimshal and like Tukur it stands on an elevated terrace. As was the case when Shipton was here, it was harvest time. Men and women were busy cutting the wheat crop with small curved sickles, carrying the bundles to threshing grounds, and threshing with teams of cattle which are driven round and round a central stake. Others again were winnowing the straw, throwing it up on the flat rooftops with long wooden forks. One small girl was threshing with a small hand flail; others were scaring the importunate flocks of red-billed choughs, which were making the most of their opportunity to stuff themselves with grain. It was a delightful rustic scene.

But there were few folk about. The Lambardar told us that the population of all three villages was about 2,000 (in contrast with the 350 mentioned by Schombergh in 1934); but we gathered that nearly all the women and some of the men, as well as most children, were still in the 'pamirs', the high pastures near the Chinese frontier, two days distant. Those we met showed great eagerness to have their photographs taken and even the women, who can seldom even be glimpsed in any Hunza village, needed little persuasion. One slightly incongruous feature of some of the groups whom we obliged was the presence of one or two transistor radios! Even in this remote place, so difficult of access, modern technology provides news and entertainment from Pakistan or China; and, we were sorry to discover later, from a Soviet news agency, presumably in Afghanistan, which relayed information about the invasion of that country in a light totally hostile to China, America and Britain.

In Tukur, we were entertained in the house of one Wali Beg, a villager of some substance. We sat around the central hearth on rugs made of yaks' wool, while tea and pancakes were prepared by Wali Beg's wife in an adjacent alcove, well equipped as a kitchen with a shining array of utensils. On one side of the square room stood a large wooden loom. As we drank our tea and picked up the pancakes from the yellow ghee which surrounded them, Wali Beg demonstrated the operation of the loom. This provides an important occupation in winter, when most people spend all their time in trying to keep warm in the bitter cold and wind. Some of the villagers remain on the 'pamirs' throughout the winter, tending the yaks, for there is not enough fodder for them in the valley. But the irrigated terraces are fertile and provide the population with a diet which seems to keep them in good health. Besides wheat, maize and barley are also grown, as well as potatoes and beans; other vegetables are little used. We saw apricot and a few mulberry and walnut trees, but a notable feature of the village, in contrast to those in the main Hunza Valley, is its relative lack of trees.

The reputed good health of this isolated settlement is remarkable; earlier travellers have reported that they had seen no goitre, nor are there signs of cretinism, although apparently the contacts with other villages are infrequent. Each family unit fetches and carries its essential needs from Pasu and Gulmit, taking their few products such as rugs and yak ghee in exchange, as and when the journey is really necessary; there is no village shop. Some of the men seek their wives in the main valley, which may well explain the lack of evidence of the inbreeding which I have seen in less remote hamlets in Baltistan. The schoolmaster, who had just returned from attending an in-service course at Gilgit, instructs the male children and we were surprised to learn that a start has recently been made, with the encouragement and assistance of the community's spiritual leader, H.H. Prince Karim, the Aga Khan, in educating the girls. The most noticeable feature of the main village is, indeed, the prayer house of these followers of the Ismaili branch of the Shia sect of Islam; an elaborate, gaudily painted building with a white roof, it attracts the eye from a distance of several kilometres.

We asked about the history of Shimshal. It appears that several centuries ago – our hosts estimated this to be nine centuries, but Schombergh claims it to have been in the seventeenth century – a shepherd of the Yeshkum caste, a Chinar speaker, came over the mountains from his native Chaprot on a hunting trip. He was a Hindu and his name was Mamur Singh. He sighted the oasis and decided to move his family there. His wife, unlike himself, was a Wakhi speaker from Gulmit and over the years this became the spoken language. Schombergh relates that the woman held her husband in some contempt and used to call him 'Shum' ('dog', in Chinar), which provides one explanation of the name. However this may be, the small family became a colony and the colony grew in prosperity to its present size and state. It was not always a state of peace, however. As I mentioned earlier, the way of life for Shimshalis has changed fairly radically within the past one hundred years, which may be taken as a reflection of political and economic changes in the outside world. In former times, we were told, the houses were concentrated into a compact group, presumably for mutual protection of the inhabitants. Today, the homesteads are scattered around the cultivated land on which their owners have their livelihood. Whereas, in the past century, gangs of Shimshalis lay up in ambush, sometimes for many days at a stretch and on short rations, to attack and loot the mercantile caravans in Turkestan, they now tend their yaks, sheep and goats on the high plateau. Unlike trends elsewhere, it would seem, life has become more tranquil, less violent.

It was time to take leave of our kind hosts. I licked my fingers to remove the sticky ghee, and after warm shaking of hands and traditional *Shukrias* (thank-you's) and expressions of *Khuda Hafiz* (God be with you) we emerged from the smoky interior to take a group photograph in the garden, before starting on our way down the hill. In the afternoon, while Robert and Jianming explored further up the left flank of the valley as far as the next glacier, Joy and I were taken, in company with a number of villagers, to see the place where the river has to be crossed by travellers to and from the pass which leads into China.

In Shipton's day a stout rope made of yak thong spanned the 150m (500ft) of water at this point; now it has been replaced by a steel hawser. But the technique employed has not changed. There was much enthusiasm among our company to demonstrate it. Joy had damaged a rib two days earlier while being carried through the river, so she declined another 'piggy-back' through the shallows as far as the near end of the rope-way. I was carried to this point but, in my turn, decided not to venture further. I had, in fact, experience of this method of crossing rivers in the Caucasus. Two men proceeded to demonstrate their skill, attaching a wooden toggle to the wire and tying themselves to this attachment for security. Then, swinging one leg over the wire and hauling themselves hand over hand, they made rapid progress to mid-stream before returning, delighted by an appreciative audience. The strength of the current and the depth and width of the river made this an impressive exercise; the more so when we realised that all the villagers, male and female, must resort to it on their way to and from the highlands. The yaks, we were informed, were driven into the water to swim, making landfalls at some distance downstream.

It was our last evening in Shimshal. We sat on the verandah of our bungalow, drinking tea and eating a special cake baked in our honour by our hosts. Daulet Amin played tunes on a banjo-like instrument; Ashraf then examined his palm and told his fortune, demonstrating yet another of his many talents. Early next morning when we started our journey back, one person was missing: it was Mahomet Rajab, the dancer. Had we been travelling in Khumbu it would have been reasonable to conclude that he was suffering a hang-over from an overnight carousal, but we were now in a land of abstinence. Rajab caught us up later and shouldered his load, which had been taken over temporarily by a local man, seeming none the worse for wear. Indeed, our men explained that he had suffered from disturbing dreams. We recalled the dance at Aliabad ten days before, when Rajab had paused from time to time to listen to the voice of the fairy, conveyed through the pipes and drum. Perhaps she had been whispering in his ear again.

I will not describe the journey down the valley, although much of the scenery, in reverse, seemed different enough. It was the more enjoyable this time because the weather smiled upon us. At the crest of the moraine above the Mulungutti Glacier, we were able to marvel at the resplendent aspect of Disteghil Sár, with the tumbling labyrinth of shining ice sweeping down towards us from the parapets of its north face. In the gorge we were able to anticipate and enjoy the rock architecture which had so over-awed us on the upward journey, and to appreciate anew the skill and labour which had created a track to surmount the many difficult and dangerous places. But we were not entirely unscathed from our experience and exertions. Joy was having considerable pain from her damaged rib; I was suffering from a severe stomach complaint, which afflicted all the members of the expedition from time to time. This had its lighter side, for Ashraf and Haidar Ali insisted on subjecting me to a 'cure' at every stopping place. Advancing on me with fell intent and with expressions which, I fancy, must identify professional torturers, they proceeded to indulge in a bout of vigorous buffeting, pummelling and massage, which included pulling my arms and legs, and even my hair! It was certainly marvellous for tired muscles, but I was unable to detect any relevant connection with the nature of my disorder.

Punctually at midday on 8 September, we again crossed the bridge over the Hunza river, above Pasu. Nigel, who had only returned that morning from escorting our Ambassador and his party all the way back to Islamabad, drove up in one of our familiar Land Rovers just as we arrived, and the vehicle was laden with 'goodies': tea and coffee and cold drinks, cakes and fruit – provender provided by the thoughtful Shane. We scarcely deserved such a generous welcome at the end of a superb trek, which we will long remember. But it made a perfect finish to our journey to Shimshal.

# The Ghulkin Glacier

It was now imperative that we carry out a new programme of radar ice-depth sounding of a glacier in the short time remaining. For this we needed a glacier with different characteristics from the Hispar, where our work would be of more practical significance. Furthermore, in this final stage of our operations, I considered it vital for us to get the Pakistan civil and military authorities more committed to our programme, since this would help them to become more appreciative of our achievements. Their involvement was doubly desirable because it was now obvious that we would never receive permission for one or two members of the British contingent to travel up the remaining 100*km* (62 miles) to the Chinese border; although we no longer wanted to include work in this direction, we did require the international character of our operations to be extended. We did not wish the authorities to realise what opportunities they had lost *after* we had returned home, but rather to grasp this last chance to collaborate with us in depth and so lay the foundations for future co-operation with scientists from other countries who might follow up our initiatives.

The fact that we did eventually achieve full-scale collaboration was due in no small measure to Major General Musthaq Ahmad Gill of the FWO of the Pakistan Army, who not only became committed to our cause but also became a personal friend. I suspect that his deeper involvement coincided with his receiving favourable reports on all our activities. About this time he appeared in Gilgit and gave me a lift in his jeep up to Hunza. During the ride he jokingly asked if I had ever been privileged to be driven by a General through such an incredible landscape as this — and along a road of which any civil engineer would be proud (he now carried overall responsibility for the Highway). Since it was raining, I asked his leave to defer my answer until the journey was completed and pointed out that, since I was now the associate of Kings and Presidents, Princes and Princesses, Lords and Ladies, Ambassadors and Ministers of State, the addition of a General or two made little difference. A smile spread across his face and in the ensuing laughter we knew that we were friends. We stopped at the roadside to share the lunch of an army maintenance crew and to discuss the problems created by recent landslides. As we drove on, I pointed out several geological and geomorphological features of the area, whilst he explained the difficulties they had experienced in building various sections of the road, so sadly evidenced by the numerous cemeteries. The consuming and insoluble problem of the Highway is the cost of its maintenance, a problem that might be eased if an operational research analysis on the effective maintenance of the road could be funded. Such an analysis would need to look at the size and frequency of landslips, the positioning, strength and composition of army or civilian units to effect repairs with speed and in safety, seasonal factors such as rainfall and freeze-thaw maxima, and the possible construction of alternative routes, possibly involving tunnels and the directing of streams *over* the roadway instead of under it, as the latter technique can undermine the Highway foundations. The bridges built by the Chinese were, of course, a source of problems of even greater dimensions. Although solidly built, they would soon require regular maintenance, since many of the abutments had to be constructed on rock that was now corroding because of the deposition of saline solutions created by the weathering of the nearby rocks.

The most serious problem was, however, the presence of the Ghulkin Glacier close to the village of Gulmit (see map, p. 100). The glacier showed signs of successive surges and retreats, a pattern not unlike the behaviour of the Hasanabad Glacier further down the valley, which is known to have advanced several kilometres in a few years. However, an advance of the Ghulkin Glacier could have far more serious consequences, since any further forward movement from its present position will destroy the Karakoram Highway; worse still, by advancing 1.6km (1 mile), it would block the entire Hunza river valley. Such an event, which is not improbable, would cause a lake to form that in turn would quickly collect millions of tons of silt, covering the Highway forever. Indeed, precisely such a catastrophe occurred a few years previously downstream, but at a place where alternative routes could be constructed.

Major General Gill drove us to the damage zone at the Ghulkin Glacier snout. There above our heads as we stood on the roadway was the threatening terminal moraine; a huge pile of rocks, boulders, stones and grit, all pushed ahead of the glacier as the solid river of ice creeps forward. About 1.6km (1 mile) wide, 1.6km long, and rising steeply from the main Hunza Valley, the moraine, the edge of which actually touches the Hunza river, is deeply cut in several places where both new and old glacier melt-water streams pour out their violent wrath.

The Karakoram Highway, with the Hunza river on one side and the glacier on the other, is forced to cross, at as low a level as possible, the creeping pile of stones and destructive, steeply cascading streams. Originally the melt-water of the glacier was concentrated into only one channel, over which a solid bridge had been built. Due to the fickle behaviour of the glacier, however, this stream had now migrated elsewhere, and the forlorn bridge stands witness to its unpredictability. The melt-stream displayed its destructive power at a new exit in the ice front by gouging out a deep channel in the moraine. Through this water-filled gully, thousands of tons of rock fragments tumble down each day to form a new fan of moraines to overlay the old. The deposition cone, clearly discernible by its contrasting bright colour, has swept away the road surface and its foundations into the Hunza river and in its place is a wide, boulder-filled stream that at the height of the midday summer melt is far too dangerous to cross either on foot or in a vehicle. Temporary bridges have been thrown across the stream, but the first was washed away and the second is temporarily withdrawn each day and resurrected every evening until a more permanent solution can be found.

Surveying the scene, we could both see and hear the stream trundling down huge boulders and in the process altering its bed and gradually creating a new route. At one point the frothing stream executed a ninety-degree turn and the force exerted at the bend was so great that we could feel the surface boulders quaking underfoot. It would only be a matter of hours before the bend would collapse and a new flow path be created, so undermining and then removing the foundations of the present temporary bridge.

Our final brief was therefore simple: to determine why and how the glacier behaved in this fashion, and to recommend actions for the future. Our first job was to study the glacier ice to see if it was of the surging type, and if so, to determine whether it was now in a state of advance and hence dangerous. This necessitated a preliminary ground survey and the acquisition of some radar ice-depth profiles. Meanwhile, the geomorphologists began an examination of the moraine and the glacier ice front.

Having just returned from Hispar, my first need was to clean up, repair kit, have a decent meal and attend to a few administrative duties. Major Rana had vanished, but I was told that he would reappear. Meanwhile, I was informed that a new Liaison Officer would soon be detailed to assist us. The next day I was in the village of Aliabad, having wooden stakes manufactured, buying coloured cloth and string and hiring the services of a sewing machinist. Within a matter of hours, and for less than £6, we had twenty flags and survey stakes for use on the Ghulkin Glacier. Arrangements were made for a contingent from the local FWO unit to help us carry up our kit, and I put in a strong request for helicopter assistance in order that we could overfly both the glacier and the moraine, so as to assess the problem from the best possible vantage point. We needed aerial photographs of the trouble spot, and fortunately Ed Derbyshire had both a ciné camera and plenty of film stock, should our request be granted. Lao Dong and I had reconnoitred the glacier by climbing through the moraines from the destroyed portion of the road. We laboriously made our way onto the glacier and carefully wandered through the ice-covered moraines in search of a site for a possible base camp. Then, at 5.00 am on 23 August, Steve Redhead drove us up to Ghulkin village in one of our British Leyland Land Rovers, a feat that defies description, and there we hired twelve local porters to supplement our detachment of six soldiers from the local FWO barracks. Marcus, Lao Dong and myself were accompanied by Robert Muir Wood.

Provisions for the porters were bought in the village, but our departure was delayed a little until we had been introduced to local dignitaries, including the Lambardar (headman) who was to be a member of a future Pakistan Everest Expedition, and the schoolmaster, who introduced us to his class of about twenty children. The environment was totally different to that of Hispar, and in contrast to the sullen attitude of those villagers, we were met by an enthusiastic commitment to our cause, a desire to share in our lives and a cheerfulness and infectious humour that mitigated against the growth of suspicion and mistrust. Everyone on our Project who eventually came to visit this village, nestling in the corner between the lateral moraine of the glacier and a steep rock buttress, concluded that it was the best village in the entire valley. It was situated well above the village of Gulmit and so was removed from the influences of the Highway, and in its isolation was manifestly an example of the kind of life that had existed in the valley for many centuries.

Twenty-two men slowly climbed up through the village to a point where the high lateral moraine squeezed itself out between ice and buttress. A short ascent brought us into a narrow, rock-strewn valley which we traversed on the right along a flank of the moraine. This pathway quickly ascended up and onto the sharp edge of the moraine and then down over loose gravel, stone, boulders and ice to the glacier itself, close by a moulin (a vertical hole in the ice) down which flowed a stream that joined others in the bowels of the glacier far below. A long but shallow rise then led up towards the centre line of the glacier.

Weaving in and out of small ice gullies and stepping over narrow crevasses, our progress was easy since our altitude had been gained in the early part of the day, when we had climbed from the village to the moraine edge. A healthy competition between porters and soldiers over who could go fastest and farthest without rests had to be firmly squashed on account of the delicate instruments many were carrying, but the happy environment of the village was transported with us all that day. Because the route we had taken was much easier than that previously used by Lao Dong and myself, it was possible to go well beyond my proposed camp site, but I made a mental note to use it as a fly camp for future survey work. Within two hours we returned to the glacier's right-hand edge, where it was now possible to ascend the moraine and traverse along a series of ablation valleys between the moraine and the valley wall. Here, shrubs grew between the huge boulders, out of a fine, sandy floor. As we climbed a little higher, there, on our right, could be seen a few remnants of seracs (ice-towers) from a once active ice-fall; then we descended into another hollow which was to become our camp for the next three weeks.

The idyllic scene beggared description. This haven was sheltered by the high moraine wall from the cool air that came from the glacier, whilst on the south side was the high east ridge wall dividing the Ghulkin and Gulmit glaciers. At the foot of the wall were old rockfalls, which I examined carefully to see if any recent slips had occurred. Far above, I noted with some apprehension a large needle of rock that appeared to sway as the backcloth of clouds drifted along the skyline. If this were to collapse . . . I diverted my attention to a few low walls which had been built by shepherds to create summer grazing grounds; a little further to the west, higher up the valley, were a few crude summer huts. But what took my breath away was the higher vista to the west, crowned by the sharp, triangular summit of Shishpare, 7,619m (25,000ft).

This peak had recently been climbed by a Polish expedition, but our porters told us that the successful summitters had subsequently walked over its edge and fallen to their death down its other flank. The summit cone stood proud of a high level of subsidiary summits, all of which channelled their snow and ice down into the long, steep beginnings of the Ghulkin Glacier. Several avalanches poured down this section of the glacier, which appeared to have more crevasses than surface. It was this incredible solid waterfall of ice, about 1.6km (1 mile) in length and height and about 0.8km (½ mile) in width, depth unknown, that constituted the reservoir of energy which would cause the lower sections of the glacier to surge forward and destroy all in front of them. The clean and shining buttresses on the flanks of the glacier, immediately below the summits, gave evidence of the extent of glacier movement and rock erosion in the area.

Amidst this dramatic evidence of the forces of nature, we made a camp on a sandy, comfortable soil, paid off all but two of our porters and set to work building a tent laboratory for all our equipment. Marcus and Robert meanwhile took the opportunity to explore the upper half of the glacier and search for the nearest water-supply.

The next day one of our porters returned to the village to acquire additional supplies of local foods, whilst Lao Dong and Marcus supervised the transportation of the electronics kit onto the glacier. I took the opportunity of placing one survey flag on a prominent ridge above camp and then traversed the glacier to place flags in suitable positions on the left bank. The crossing was easy, but the thought crossed my mind that even though we were now short of time, perhaps it was unwise to have set out on my own, since no-one knew where I would have to climb to find intervisible survey stations. The first two sites were about 100$m$ (330$ft$) above the level of our camp site, but after siting a third flag downstream on the lip of a crumbling moraine edge – not suitable for a station, but excellent for a triangulation resection with other stations – it became clear that I should replace my second station about 250$m$ (820$ft$) above the glacier. Here there was a tiny platform on a sharply descending rock ridge which would command far better views upstream, downstream and over the entire middle half of the glacier, as well as being clearly positioned for the further stations I proposed for the other side of the glacier. By the time I had modified the second station and enjoyed a low-grade rock climb on loose rock, I retired to the shade of a boulder to prepare two more flags.

The major station was to be located upstream, as close as possible to the steep ice-fall on the left-hand side of the glacier and high on a rock rib. The walk up the ablation valley was gentle, and for the first time for many days I felt as though I could relax. Frequently I stopped to look back to examine the ever-changing detail, which in fact made it difficult to relocate the flag pole on the now-receding rock rib. Twice it disappeared from view, but I knew exactly where to look when I had climbed a little higher. Nonetheless, if it were lost and mistakenly relocated, this would cause confusion when we later came to establish the triangles between all the stations – a fundamental principle of the art of triangulation and mapping.[1]

After an hour's walk I came to a steeply descending scree in a gully composed only of large clean-looking blocks covered with dust, evidence of a recent major rockfall that had landed straight on the glacier. On the far side of the gully, and above me, were two small trees growing out of an alcove which promised a short and protected climb onto the bounding rib. I was pleased to reach the trees, since several of the large rock blocks that had fallen in the gully were precariously balanced. Now I saw that I had to climb a short overhanging crack to gain access to the rib. Fortunately the rock rib, although steep, was not too loose and soon I was planting a flag in an excellent position that commanded a very wide view. I just had to hope that the distance was not too great to permit me to see this flag from the mountain wall behind our Ghulkin base camp.

Carefully and hesitantly I retraced my steps, partly because of the loose and steep rock and partly because I was alone, but within the hour I was back in the ablation valley, and relieved to be on my way home. Another hour passed before I reached the shade of my friendly boulder.

Recrossing the glacier, I made a detour to the radar team, only to discover that the Vatnajökull equipment seemed to have blown a fuse and that the new equipment was giving no response – although the ice here was presumably shallower than that of the Hispar Glacier. Fortunately it was the fuse that was at fault, and this was soon replaced.

We were now almost ready to start, but before an integrated programme of survey and ice-depth sounding could be planned, I still needed a further two stations upstream on the right flank of the glacier. On 25 August I went up the continuation of the ablation valley ahead of our snug camp site, and soon located two further stations. It was essential that all other flags could be seen from the first station, and since the glacier now curved to the south before climbing steeply up the major ice-fall, the penultimate station was not a great distance from the camp. The last station was on a 3m (10ft) high boulder, flat-topped and possibly precariously balanced on the edge of the moraine. Three rock climbs of 'difficult', 'very difficult' and 'severe' standard, one on each face of the boulder, distracted me a while from the task in hand and gave me some enjoyment, but I eventually proceeded to build a small cairn and plant a flag pole. I then jumped up and down to make sure the rock was stable. Thankfully it was. Unfortunately, from this last station only three of the other five flags could be seen, the minimum number required for a survey resection. I now retreated to start the theodolite survey from each of the flags. It was only then that I discovered that the theodolite we had brought up did not have a sufficiently powerful telescope to sight the flags, some of which were separated by at least 5km (3 miles).

Fortunately my return to Aliabad in the very early hours of the next day to acquire a larger theodolite allowed me to greet Lord and Lady Hunt, who had arrived in Hunza for a holiday. I think I may have possibly disappointed a few people by not being present on the day of their arrival, but I had anticipated that once John and Joy knew of the urgency of our work on the Ghulkin, they would not think twice about the issue; indeed, if the positions had been reversed, I felt justified in assuming that they would have done exactly the same in the interests of the expedition. Nevertheless, I have to admit to a certain degree of apprehension about the meeting, since John had a commanding personality and presence and had not yet been fully briefed about our current position. Tall, strong, apparently ageless at seventy and with a determined character, he can bring so much experience to bear on almost any topic that he is eagerly sought-after for all kinds of operations. However, I already knew we were two different personalities with possibly different approaches to the job of leading large expeditions. He had led the successful Everest 1953 expedition by being in the van, and correctly so. By contrast, I was leading a scientific expedition from within, delegating duties to those who could do specific jobs far better than I could. Furthermore, academics have a totally different attitude to that of mountaineers – an attitude which often prevents them acquiring the mountaineering philosophy so essential to the ascent of big mountains. But I knew that John's overriding intention was to escape the trappings of civilised life, if only for a few days, a sentiment I could fully appreciate. So when I arrived I was pleased to see that he was already the centre of attention around the breakfast table, busily conducting discussions with members of our expedition at base camp.

That day we quickly decided on a three-week programme. With not a little envy, I suspected that John and Joy would wish to visit Shimshal, a place of great historic interest, as well as being the most remote village of the Karakoram and hence possibly the world. I was not disappointed, and only wished I could have had the time to accompany them.

Within the next few days I quickly realised that John's determined and aggressive nature was self-imposed, and was strictly for the purposes of self-motivation; whilst these are the very qualities required to overcome the difficulties of mountain terrain, they were never a characteristic of his relationships with other people. 'Quiet', 'unassuming', 'attentive', 'perceptive', 'energetic' and 'encouraging' were the various terms used by our members to express their appreciation of the two new arrivals. Indeed, he and Joy fitted into the harmonious spirit of our venture immediately, and made a point of seeing all the members and asking questions on their work programmes. Within the day they became fully integrated members of our expedition. It was what we had hoped for, and what they had desired.

The loose, but loyal bonds between groups and members of the expedition, which had grown in strength during the course of our shared experience, were further exemplified when I asked Jon Walton and Tom Crompton if they could forego their holiday and bring tellurometers (distance-measuring devices) up to the Ghulkin to help us complete our work. They readily agreed to do so and later were also prepared to do river-bed level measurements for the geomorphological programme. However, neither of these tasks could be carried out until early September, and it was essential that I return quickly to the Ghulkin camp the next day, if only to acquire an initial round of theodolite angle readings. Late that night, pleasantly tired, I wandered into the wrong tent (there were so many at the Aliabad base) and tried to climb into a sleeping bag already occupied by Jianming. His quiet protestations indicated that surveyors may be friends, but international relationships should not be strained.

On 27 August I returned to Ghulkin with Joy and John and Bob Holmes. We had unanimously agreed that Joy and John should have a quiet three-day trip up into the mountains to help them acclimatise, although they soon proved that we need not have been so concerned. While they started up the track, I acquired more supplies of atta (flour), ghee (butter), potatoes and cigarettes for our porters and then, carefully handling the delicate but heavy theodolite, I set off once more for our beautiful camp site. John was quite poorly with the usual stomach bug, but he and Joy, with the measured paces of experienced mountaineers, covered the distance to camp in a surprisingly short time. My education concerning the fitness and abilities of British peers was about to be completed. John's fluent Urdu was very much appreciated by our porters, and Joy's willingness to take over the kitchen was gratefully accepted.

Soon after arriving in camp, Bob and I went onto the glacier to locate Marcus and Lao Dong, while John and Joy went upstream to visit our stations. On their return I was surprised to learn that John had ascended the boulder station by two of the hardest routes, one of which required an agile move onto a mantleshelf (a high narrow ledge) before the top of the boulder could be grasped.

The following day we all ascended one of the survey stations and a full round of readings was taken, four of the flags now being readily observable through the large telescope of the new theodolite. Whilst I visited other stations, everyone went onto the glacier to examine the radar kit, which was now functioning perfectly. Marcus and Lao Dong were keen to get all the results they could acquire on each day that remained, and in order to assist the precise location of these depth-soundings, they built cairns on which we could later set up the theodolite and observe our flags on each of the glacier flanks.

The following day, Joy, John and Bob descended back to Ghulkin and thence to Aliabad, while I returned to camp. That evening we compared notes and decided on our final strategy for the completion of the glacier survey, which would be much simpler now that Jon and Tom could bring their tellurometers. Marcus and Lao Dong would continue taking ice-depth soundings down the centre line of the glacier and complete the second of three proposed transverse profiles. Both of these scientists had worked hard and long and no doubt were now looking forward to a few days' holiday away from the rigours of scientific research.

Early the following morning I once again left for Aliabad. I departed somewhat reluctantly, since I had enjoyed my few days on the glacier – especially as it now appeared that all programmes would be completed satisfactorily, with far more data acquired than had been thought possible during the planning stages. My only worry about the Ghulkin survey was whether or not we would be able to relocate the cairns left by Lao Dong and Marcus among the moraines which covered the glacier. The major relocation problem would occur downstream close to the snout, where the ice surface and moraine cover was most uneven and one moraine mound looked very much like the next only a few metres away.

Preliminary depth-sounding results showed that the ice was about 400$m$ (1,300$ft$) thick in the middle and that it maintained this thickness almost to the snout, where thinning took place. No evidence of surging could be found; the ice was wasting in the middle, and the ice-surface level was well below the height of the lateral moraines, which were rapidly disintegrating. Crevassing in the lower half of the glacier was almost non-existent and an old ice-fall had almost ablated away. The Ghulkin was therefore clearly not advancing and the present problem obviously lay in the snout itself. This had the appearance of splayed fingers, representing five different ice-flows. Old melt-streams had issued from different zones at various times and the recent change was probably due to the collapse of an underground ice tunnel, which had caused the under-ice melt-stream to be diverted. Unfortunately such behaviour occurs regularly and will continue to do so.

Similar sequences of events related to changing glacier conditions can be observed on neighbouring glaciers, but they are seldom noted, since they are far removed from critical zones. It is only because the Ghulkin Glacier has its snout so close to the Highway and the Hunza river that sudden changes cannot be naturally accommodated without affecting man's works and daily life.

Back in Aliabad I was delighted when Jon and Tom asked if it would be possible for me to join them on the Ghulkin, since they now knew the extent of our work and wished to complete the task as quickly as possible with as many hands as could be spared. Before this, however, we all joined in an event that was to become known as the Hunza Sports Day.

Nigel had taken up my earlier idea of an international football or cricket match and had brilliantly and imaginatively expanded it to fit local conditions and to involve all the different ethnic groups in the locality. Furthermore, he had acquired a silver cup, suitably engraved for the occasion. Four teams were to compete, the challengers being the IKP 1980 expedition and the challenged the Hunzakuts' A and B teams and the FWO. Fortunately the Hunzakuts, always a strong and competitive race, are now more attuned to western ideas of sport than their warlike grandfathers.

All the games involved some degree of strength and skill, but since they were derived from local sports, we thought we had little chance to shine. The first competition was throwing the stone (shot) from a stone-ringed circle, but despite the beautiful attire of our team, who were wearing lilywhite tee-shirts edged with blue and proclaiming 'Chloride batteries last longer and give more energy', all our throw markers were at least a metre behind the main bunch, some 14m (46ft) from the stone circle.

The next competition was one-legged fighting, a popular game in the Karakoram valleys and one that I had played on previous expeditions. One leg is gripped behind one's back and, whilst you hop around in a closed circle, your free hand is used to grapple with and topple over your rival. Going for a rapid surprise victory I made too quick a lurch and my wily opponent side-hopped and helped me on the way with a gentle push. My vanity evaporated as I hit the dust. But within a few minutes the IKP were cheering at last as Ted Smith won his round to go into the quarter-finals. He was our only victor of the day, although we almost caught the Hunza tug-of-war team unawares. This game is now their second most popular local sport, after polo, and being tall, strong and well-built, it came as no surprise to us to learn that they were the champions of the Gilgit tournament. During a pull they execute an unusual manoeuvre which involves the front man turning round on his opponents and pulling forwards, to be quickly followed by each member in turn. Alternatively, if the opponents show obvious signs of weakening, the team will turn in unison and with the rope across their backs simply drag the losers through clouds of dust into ignominy. But our team had an expert in Denys Brunsden, who taught his stalwarts never to let the rope be pulled high enough for the Hunzakut team to perform this trick, which would signify to all present that defeat was nigh. The IKP team not only held the Hunzakuts at bay longer than any other team, but at one instant regained lost ground before losing, and so honour was satisfied.

That day the entire male population of Aliabad, if not Hunza, squeezed into the

Land Rover crossing
the treacherous
Ghulkin melt-water
stream.

Ron Charlesworth
filming in the Hunza
valley.

base camp to applaud, laugh, cheer and enter into the spirit of the competition, at one point waiting with patience and understanding until the referees had adjudicated on whether or not the FWO team could wear boots in the tug-of-war. After long discussions, the barefooted Hunzakuts were allowed to keep on their woollen overcoats to help increase the frictional force of the rope, whilst the FWO team were dragged, boots and all, across the line, to resounding shouts from the crowds.

Lord and Lady Hunt and local dignitaries sat enclosed within a marquee to keep an enthralled and not too dignified watch on proceedings. Shane and Helen kept the record sheets, whilst everyone else was madly shooting film of an event which will long be remembered by all who attended. For the evening celebrations, Nigel and Shane had prepared a banquet for seventy guests, but all one hundred and twenty attendees were amply fed from two barbecued sheep. The day was brought to a close by the villagers giving us a taste of their own style of entertainment. Several sketches had been prepared, including parodies of porters assisting sahibs to the summits of peaks, hunting ibex, and doctors examining reluctant patients, all of which created great amusement. The evening was brought to a close by sword-dancing and a group of children performing local and Punjabi traditional dances, to the enthusiastic delight of all present. At last, when all the prizes won during the day had been presented, the team leader of the Hunzakuts gave a celebration dance, and I could not help but reflect on all the changes that had occurred here in the last eighty years. These people were now our firm friends. Younghusband and Shipton would have been proud to have been in our company, Durand would have been perplexed, the Maharaja of Kashmir disappointed, but Gardiner and Leitner would doubtless have been thankful to have seen how young Nigel and Shane had helped cement our international friendships (see Appendix I).

My lasting memory of the day was Tom Crompton attempting the Limbo dance, clad in baggy shorts and exhibiting his usual broad, infectious smile, whilst Denys Brunsden tapped the drunken heads of the local policemen with batons, only slightly less intensely than they had done during the day to keep all the young boys back into the pressing crowds. It was only the rain that eventually sent us back to our tents, calling a close to a most remarkable day.

Before journeying back to our now uninhabited Ghulkin glacier camp – Lao Dong and Marcus having returned to Aliabad in time for the sports day – Joy, John, Helen and I visited Prince Ghazanfar Ali Khan, Queen Rani Jamal Khan, and Princesses of Hunza for a most delightful dinner party. During the evening I discussed the women's liberation movement with Begum Ali Khan and Princess Azra Sher Khan, and was pleased to note the rapid advances being made; but being aware of the different attitudes of neighbouring Muslim sects and the wide range of social and educational levels in any one community, I knew these advances were only relative and that the womenfolk still had a long struggle ahead to overcome the inequality gap. We also talked of developments in medical and educational fields and other needs in this area, where population expansion is creating a desperate need for new and safe land for cultivation.

Before leaving the Palace, I looked at the pictures and gifts decorating the panelled walls of this fine home, many of which had been presented to the Mirs of Hunza by members of the British government and the Royal Household. Honours such as Knight of the British Empire and Commander of the Indian Empire had been conferred on past Mirs, and there were numerous photographs of them beside various dignitaries of the British Raj, including Governor Generals and Kings, particularly George V and VI taken at Royal Durbars (festivals). It was perfectly clear that these photographs and decorations were kept permanently in position and that their owners had a genuine regard for the Raj and what it had done to assist the government of the people of both Hunza and Nagar.

During this brief return to Aliabad I also attended a dinner party with Major General Gill and his FWO staff. Once again helicopter assistance was requested and general agreement reached that it would be helpful to everyone if we could take home the Pakistan Air Force aerial photographs for a more detailed study, to assist the assessment of hazards on the Karakoram Highway. Meanwhile John and Joy had rushed up and down the hillsides of Hachindar, 4,544m (14,900ft), talked to all Directors of field programmes and most of our members still working in the Hunza Gorge, as well as preparing memoranda for discussions with VIPs in Islamabad concerning our work.

The end of the Project was now fast approaching. One group led by John Hunt was en route for Shimshal, and I, Jon and Tom were on our way to the Ghulkin Glacier. Torrential rains stopped us doing any work on 2 September, except for the repositioning of one fallen flag, and the examination of three other sites to ensure all flags were in good order, but the following day, Jon and I went off to the shepherds' huts and completed our first telemetric station via Tom, now positioned on the rock rib above our camp. From here we completed the upper transverse section of the glacier and climbed the far rib close to rockfall gully, but by a much safer route on its western side.

During the traverse I espied a most remarkable phenomenon; water spurting out like a geyser from a shallow crevasse. At first I could hardly believe my eyes; but I then took a bet (a beer) that the event was real and would probably be repeated. It did, and this time I felt and tasted the water, which had emerged in a 1.5m (5ft) spout. After my experiences on Vatnajökull, I had no idea what to expect; in fact the water turned out not to have a sulphurous taste, nor was it hot. It was probably being ejected because a below-surface melt-stream channel was being compressed as the ice moved forward.

That day we visited many cairns. All were located with ease except two, which we managed to locate precisely from relative data on pacing and timing intervals between neighbouring stations. We trudged back to camp in the late evening to meet Tom, who had by then surveyed from three of my flags and had had clear views on all others except one, which I had planted on the very edge of the moraine just above the village. The evening meal was most enjoyable. Above us, Shishpare turned white, yellow, orange, gold, and finally silver as the sun sank, long after our hidden valley had succumbed to dark shadows. The following day, Jon, Tom and a porter would complete the survey, since we had, in one single day, taken readings at more than half of the ice-depth stations marked by the cairns built by Lao Dong and Marcus.

The next morning, somewhat sadly, I walked back down to the village with our porter, taking with me the tellurometer reflector and a survey pole.

The flag previously planted at the bottom of the lateral moraine had been stolen, so I sited another and then tuned in to Tom to enable him to measure the separation distance between us. The porter then returned to camp with the equipment, while I, for the very last time, walked down the steep track into the village, pausing awhile in the rock-filled valley to see if I could catch sight of what I thought had been a fox on the way up, several days previously.

A small party from the British Embassy awaited me, and that day we were able to repay in some small measure all the effort and goodwill bestowed upon our expedition by the Embassy staff. After an examination of ancient rock scripts being studied by a German expedition, we chose to have lunch by the river, underneath, or so it seemed, the Tolkien-like peaks guarding the flanks of the Shimshal Valley. Oliver Forster the Ambassador, Arlene and Bill Fullerton, Brenda Wheeler, Nigel, Shane and myself could all now relax for a few hours.

A day later, I was back in Gilgit. However, about this time a helicopter had been made available for an aerial survey of the capricious Ghulkin Glacier, and Ed Derbyshire and Marcus Francis were busy shooting film that successfully completed our work on that glacier.

Bad weather prevented us flying back to Islamabad and since time was now short, we had no option but to take to the Karakoram Highway. Apart from a few lectures in Islamabad, the writing and presentation of reports to various Ministries and receptions given by our Chinese, British and Pakistani friends, the expedition was over.

# Conclusion

This story has been based on the record of only one member of the expedition, perhaps one better placed than most to see the whole adventure in perspective, but nevertheless one who cannot do justice in so few pages to all that was achieved by the seventy or so members of the programme.

In reality, this story has no beginning and no end, but is only a part of man's continuing endeavour to understand and surmount the difficulties of his environment. It is within that context that we must place the tragic event of 14 July. Undoubtedly greater accidents have occurred in the Karakoram in the past, and more will occur in the future to shake our faith; but it is some consolation to me and to my friends that Jim Bishop was participating in an endeavour that brought together the scientists of several countries, notably from China, Pakistan, Britain and Switzerland, but also including assistance from scientists in Italy, Germany and America. The accident will always tinge with sadness the memories of every person who participated, directly or indirectly, in the venture. However, having worked with Jim for many years, I know he would be pleased that, as a lasting memorial, his colleagues, friends and relatives have set up a trust fund in his name at the Royal Geographical Society, to assist young men and women to go out and examine the wild and beautiful places that he so loved during his all too short life. A more fitting tribute could not have been designed.

The Karakoram Project 1980 initiated a series of complementary scientific studies which enabled us to all experience what could be gained from collaborative learning and international co-operation. Introducing technology into all man's activities cannot, and should not, be avoided if we are to build a better world, but it is essential that first we all fully appreciate what other disciplines, and other cultures, can contribute. Only by this means will we eventually convince governments that despite our cultural, racial, political, social and economic differences, we have one common concern; the protection of our environment. Where better to initiate this ideology than in the Karakoram, with a group of international scientists using modern technology to try to unravel the mysteries of nature?

It was therefore unfortunate that we were unable to convince political journalists in the USSR that our venture was purely scientific. On 5 August the press agency Agentstvo Pressy Novosti of Moscow sent out a news release which was duly reported the following day in the Indian press. It stated that our team included 'two Americans from the CIA and the National Security Agency . . . an Englishman, Miller, who is an "expert" in clearing hurdles on other people's borders, as well as Chinese and Pakistani intelligence personnel.' It continued: 'The American experts are entrusted with the task of choosing a site for computers to gather intelligence data from the Soviet Union, Afghanistan and India. The "mountaineers" expedition

[is] code-named Karakoram 80 . . . .'

Similar articles appeared in other leading Indian and East European newspapers, and eventually, to close the loop, *Pravda* reported on 7 September that 'The foreign press reports . . . a strange "expedition" [which] clearly has no relationship with science [but] in the most direct way is connected with the American CIA.' The remainder of the article relayed equally misleading information based on the original 5 August APN news release. It gives me some satisfaction, however, to know that at least four large colour photographs of our expedition adorn the houses of Russian colleagues. When next I lecture in Moscow, I shall also recommend our kind of international venture to the scientific societies of the USSR.

With regard to our fieldwork, I now realise in retrospect that we aimed to do too much; surprisingly, however, we achieved even more than we ever thought possible. This was due in large measure to friends in Pakistan, in particular to David and Alison Latter of the British Council, who encouraged their British and Pakistani friends to join the vast army of our supporters, and also to Oliver and Beryl Forster of the British Embassy, who together with their Pakistani and British staff provided substantial commitment to our cause. Lord Shackleton and Lord Hunt also made positive contributions and resisted the temptation to interfere even when they had a right to do so, understanding the mammoth problems to be tackled by each and every Director of our programmes. In Pakistan, men of outstanding integrity such as Dr Afzal, Dr Shami, Dr Manzoor Ahmed Sheikh and their respective teams served us with pride, discretion and strong resolution. They, more than anyone, knew what we were up against and quietly and patiently paved our way to success, along a road built and navigated with skill by our friends in the FWO of Pakistan. We shall forever be in debt to the military authorities led by Major General Gill and, before him, by Major General Butt. We hope also that eventually, when the results of our data have been computed, they in turn will be in *our* debt, and that the results from our fieldwork will be of practical relevance and importance to Pakistan.

Strong links have been made with civil, military and educational establishments in Pakistan concerned with work related to the safety of people living in hazardous environments. It is hoped that historic buildings in the Karakoram can be protected, maintained and preserved by international societies, some of whom helped finance our project. One important outcome of the interdisciplinary nature of our programmes was that the IKP highlighted the necessity for improvements in the medical, technical and educational facilities available to the people of the Karakoram. For example, villages in the alpine belt require fire-fighting equipment to protect their wooden houses – a far cheaper facility than providing aid after an earthquake. Likewise, the builders in expanding villages in timber-scarce zones need to be taught the uses and misuses of new material such as reinforced concrete. The seismologists' work may never assist these people directly, but this project in particular will help bring together other scientists of similar international repute; by working in these earthquake zones, all will come to appreciate the need for large-scale international collaboration across traditional discipline boundaries. The surveyors made a positive contribution to all programmes as well as their own, and with their advanced instruments saved scores of man-years of effort by the Survey of Pakistan, relocating lost or misplaced survey stations, and establishing several new, permanent stations via the satellite receiver.

Nigel and Shane, and our doctors, David and Helen, backed by Bob Stoodley and Steve Redhead, not only kept us constantly fed and mobile, but cared for us and the people we associated with to the very limits of their energy. Professor Tahirkheli and Qasim Jan spearheaded the Pakistani contribution, and in their wake brought the valuable support of the staff of the Geological Survey of Pakistan, who should now have no doubts or fears about future collaboration with similar projects. The glaciological programme, although severely limited by the rapid depletion of its membership, was a particular triumph. We acquired all the results we desired, due mainly to the unceasing efforts of Marcus Francis, Gordon Oswald and Lao Dong. The wonderful, last-minute present we received when a helicopter arrived to allow Ed Derbyshire and Marcus Francis to take an aerial ciné film of the Ghulkin Glacier and so enable a more comprehensive analysis to be undertaken will be forever in my mind.

Perhaps one of the more long-lasting memorials to the expedition will be the two volumes of proceedings of technical papers (see Appendix IV, p. 200) presented by expedition members at two conferences, one at Quaid-i-Azam University before the start of our fieldwork, and one at the Royal Geographial Society in September 1981. These volumes, together with the formation of the Karakoram Research Cell under the guiding influence of Dr Afzal, should be of great benefit to all future scientists visiting Pakistan and hoping to collaborate as equal partners with the academics of that country. It is hoped that the second international conference on Recent Technological Developments in Earth Sciences in Islamabad in 1985 will be sponsored by the Karakoram Research Cell.

No words can express our appreciation of the Chinese contribution to our Project. Our friendship started quietly and almost reverently in the tents of Aliabad, but the Chinese quickly shed their initial reserve and displayed an infectious enthusiasm for work which, coupled with a well-disciplined approach to research and exacting analysis, cemented that friendship for life. We were very sorry indeed to lose their sincere and cheerful comradeship when the time for parting arrived.

Expeditions also depend for their success on the activities of countless supporters back home. At times of difficulty, I only had to remember the smile of John Hemming, the rigid support of John Auden and the assistance of numerous secretaries. Our success was also a reflection of the number of scientific grants in aid received by the expedition from the most discerning research committees in Britain, namely, the Royal Society, the Science and Engineering Research Council, the Natural Environment Research Council and the Overseas Development Administration. These, together with grants from the Royal Geographical Society, the Mount Everest Foundation and the Royal Institution of Chartered Surveyors, gave us most of the financial support we required from the United Kingdom. The National Geographic Society of the USA gave us a grant that permitted the vast number of geomorphological studies to be conducted; ably directed by the Deputy Leader of the Project, Andrew Goudie, this group completed the most extensive land-form study yet attempted in the Hunza Gorge, and so gave valuable support and information to the Water and Power Development Authority of Lahore, who kindly assisted by providing staff and a jeep to despatch teams all over the area. British Leyland loaned us four Land Rovers that never once broke down, a tribute to the skills of British workmen that should never be underestimated.

Before we flew home by British Airways – who co-ordinated our sometimes complex movements, involving tons of baggage and highly delicate and sophisticated equipment – there occurred one final incident that neatly summarises the difficulties and frustrations to be faced by any expedition that has to deal with an understaffed and nervous bureaucracy. The aerial photographs of the Hunza Gorge had been handed over to my care with a clear warning that they should not leave the country, be returned at the end of the expedition without loss, always be produced on demand, and under no circumstances be used for any purpose to the detriment of Pakistan. I agreed to these conditions, despite the fact that we had had to pay £300 to procure them and that we would have been able to extract more data to the benefit of Pakistan if we had been permitted to retain them. As the day of departure drew near, it became increasingly obvious to many people with whom we liaised that it would be better if we kept the photos for further analysis back in Britain, but nevertheless I took them back to the Survey of Pakistan three days before my departure. There, to my surprise, I was requested to retain the photos, despite the now ominous warnings of the Ministry official who was our main link-man with all government departments. This extraordinary impasse could not be resolved. No one appeared willing to take the photographs back – not even the official who insisted that I should keep to my bargain, maintaining that because they were not his property, he could not check them. I therefore decided to leave the photos with the British Council acting representative, Richard Hale, with instructions to hand them over after I had left the country.

Before boarding the aircraft, however, I was drawn out of our party by two officials and asked to sign papers to absolve them from any possible trouble. Perhaps they thought I was trying to smuggle the photographs out of Pakistan. Not wishing to create any further problems, I hastily telephoned Major General Gill to inform him that some bureaucratic difficulty threatened to delay my departure. With only a few minutes before the plane took off, and with all members of the party already on board, Major Tahir drove up to the VIP lounge at speed, and with two military officers on either side, I was safely escorted to the aircraft, leaving my inquisitioners staring disconsolately after me.

As an engineer, I also derived great pleasure from the fact that so many industrial and commercial organisations were involved (see Appendix III, p. 197). By far the greatest contribution came from George Wimpey Limited, who sponsored the survey programme; not only did they meet half the cost of this work, but they also took an active interest in all aspects of the study, including the analysis of the results. I hope that in future other large engineering organisations will become involved in similar projects, as these can very effectively transfer the abilities and expertise of advanced countries to those who have fewer highly-trained scientists but who are eager to acquire our skills and to increase their own rate of development.

And so our field operations came to an end, although the analysis of our data will take many more months to complete. The continuation of this story will again depend on the constant, cheerful and unstinting collaboration of all members, whose most endearing quality was their unreserved friendship – the hallmark of the International Karakoram Project, 1980.

But the final words must be addressed to the Pakistan Government, who must be thanked for allowing us to carry out our extensive range of studies in an area that has generated as many suspicions as natural calamities. Unquestionably, we could have caused considerable embarrassment if anything untoward had happened. Their faith in us was our reward.

# Appendix I: History of Exploration

*Early Exploration*

The first recorded travels in the mountains surrounding the Karakoram took place many years after Buddhist monks had penetrated into India from the north during the fifth, sixth and seventh centuries AD. The renowned Marco Polo provided a few details of a journey across the Pamirs from Balkh to Kashgar in 1274, but this did not constitute a crossing of the mighty Karakoram, which even today few have achieved. Before 1800 a Jesuit missionary had crossed the Himalaya via Ladakh as far as Lhasa, and one British expedition had gone to Kabul to examine the possibility of a French invasion. This expedition produced a map which, although inaccurate, recorded the names of the 'Kurrakooram' and the 'Hemalleh' mountains and it also noted the true direction of the flow of the Indus. Even earlier, in 1663, Monsieur Bernier had travelled as far as the Vale of Kashmir, and Kashmir had also been visited in 1783 by an Englishman, George Forster, en route to Britain from Bengal via Afghanistan and Russia.

The first exploration of note was that of William Moorcroft, who in 1812 crossed the central Himalaya via the Niti Pass of Garhwal and reached Gartok. In so doing he not only established the source of the Sutlej, which had previously been thought of as the birthplace of the Ganges, but he also correctly identified another river to the north as the Indus, although it flowed north-west into Little Tibet, an incredible judgement for that time (see Figs 1 and 2, pp. 1, 4, and 5).

On a second great journey with an Englishman called Trebeck, Moorcroft came into conflict with the local tribes and in particular with the Maharaja of the Punjab, Ranjit Singh, to whom he had to report. When only eighteen years of age, this one-eyed, illiterate but dynamic leader had captured Lahore and united the Sikh leaders to repel the Afghans. Moorcroft was kept under arrest in atrocious conditions for the whole of April. It was not until August that he was able to lead his large caravan across the higher passes to the north, after being grudgingly granted permission by the Maharaja. It must be remembered that at this time the boundaries of British influence exercised by the East India Company did not go further than Bilaspur on the Sutlej, and so Moorcroft was regarded as a trespassing foreigner and was expected to pay tributes to all minor and major officials of every state, big and small, through which he passed. In the meantime, however, he enjoyed the attentions of the local women, even though he was then fifty-five years old, a relatively advanced age at that time, bearing in mind the climate and diseases of the countries through which he travelled.

During the winter of 1820–21 Moorcroft rested in Ladakh, but applied to the Chinese authorities, *640km* (400 miles) away in Yarkand, for permission to proceed across the Karakoram. He waited three years. The Kashmiris themselves were also unsympathetic, since they controlled the trade on the southern flanks of the Karakoram and, like the Chinese, saw no reason to upset the status quo. Fifty years were to pass before a Briton succeeded in crossing the more hazardous passes of the Karakoram and the Kun Lun.

During his prolonged wait in Ladakh, Moorcroft came to love the hospitable peoples of this inhospitable land, but when he tried to further their petition for British protection he fell foul of the political alliance between Delhi and Lahore and only just escaped to Afghanistan via Kashmir and Peshawar before appeals for his return were despatched. Passing through Kabul, he then crossed the Hindu Kush into Badakshan (north-east Afghanistan) via the Hajigak Pass, arriving in Khulm in September 1824. Here he was held captive by Murad Beg, the brigand chief of Kunduz, and his expedition began to disintegrate. He eventually achieved release and five years after starting out he came to the end of his journey, Bukhara in Western Turkestan.

By the time Moorcroft set out on his return journey through the territories dominated by Murad Beg, much suspicion and mutual distrust had been generated. Within a few weeks, the entire expedition was annihilated and its property spread throughout Afghanistan. The Royal Geographical Society, founded in 1830, published some details of these journeys, but little could be discovered about the deaths of Moorcroft and his companions. Whether they died from fever, poison, or assassination will never be known, but from reports by later travellers it would appear likely that they fell prey to brigands. Perhaps it is worth noting that when Jim Bishop, a member of our Project, visited Badakshan in 1971, he and his friends were also robbed of all their goods, including the scientific results painstakingly accumulated during the previous two months. Fortunately, however, they did not awaken or stir while being stripped of their possessions.

In the early years of exploration, the centre of political and geographical importance for travellers wanting access to the unexplored lands to the north was Lahore, where permits could be obtained. Thus it was fortunate that in 1831 a French botanist, Victor Vinceslas Jacquemont, was befriended by the Maharaja of Lahore, Ranjit Singh. Jacquemont travelled to Kashmir with the latter's permission, although this by no means guaranteed him an easy passage since he was temporarily arrested en route by Ahmed Shah, Raja of Baltistan. Because of his friendship with Ranjit Singh, who was still keeping a surreptitious eye on his activities, and no doubt wishing to proclaim his allegiance to Lahore, Jacquemont denounced the overtures made to him by his captor who, mistaking him for an Englishman, requested the support of the British! Perhaps understandably, two years later the Sikhs started the first of several attempts to invade Baltistan.

All explorers in the Karakoram tend to be individualists. Moorcroft was an unquenchable enthusiast looking for commercial opportunities, Jacquemont a sophisticated dilettante; but hot on their heels followed a religious misfit hoping to promote Christianity with the Jews of the Middle East, a man called Joseph Wolff. Wolff travelled east through Afghanistan to Bukhara, avoided Murad Beg at Khulm and eventually arrived at Doab, only to be arrested – not for the first time – by brigands and brought before the Mullahs (priests). Stripped of his few worldly goods, including all his clothes, he escaped execution at the hands of the Hazara tribe by pronouncing that the Koran instructed that infidels be respected. He was wise enough to give the name of Murad Beg as his protector and when set free, headed south before his frightened captors could check his story. At Kabul he was rescued by Colonel Sir Alexander Burnes (formerly a Lieutenant in the Bombay army) en route for Bukhara – a most fortunate and incredible coincidence, which saved his life. Soon he was off again to Kashmir.

Treated well by Ranjit Singh in Lahore, presumably because he was evidently a holy man, he went on unhindered to Simla and from there to Kashmir. Perhaps it is a testimony to the difficulty of the terrain that this experienced traveller, who had journeyed across some of the most difficult country to the east, made no attempt to cross the high passes to the north in the late month of October and instead retreated to India.

In 1835 another Englishman, Geoffrey Thomas Vigne, with no motive other than pure inquisitiveness and a desire to seek cooler climes, crossed the Sutlej at Bilaspur, as Moorcroft had done some fifteen years previously. A more pragmatic and accommodating explorer than his predecessors, Vigne was a keen hunter and so had a common understanding with those with whom he had to deal. His other talent was drawing, and he was able to please Ranjit Singh by immortalising the Maharaja on canvas. 'Ramrod', as he was commonly known, crossed the Himalaya six times from Kashmir, and five times crossed the Pir Panjal. Like Moorcroft and Trebeck, he was of the opinion that Kashmir should be annexed as a part of India. Although this area did eventually come under British protection, it never became a part of the British Raj. In fact its fate was not decided until 1948, and even today the matter is still in dispute.

As soon as Vigne reached Srinagar, he headed north over the Himalaya and became probably the first European to see and report on what was then thought to be the second highest mountain in the world, Nanga Parbat, 8,125m (26,660ft), standing aloof and threatening almost 80km (50 miles) away to the north. At the Burzil Pass he was met by emissaries of the Raja of Baltistan, Ahmed Shah, who was still keen to have a British ally. Because of conflicts of interest between the Sikhs, the Baltis, the British and himself, suspicions were sown that were to bedevil expeditions in the area for many years to come. It must be remembered that zones of influence in these high, frequently impenetrable and harsh lands are seldom more than a collection of villages which frequently change their allegiance. Then, as now, there was no clearly definable border between states, provinces or countries. Numerous stories record how brother would slay brother, mother or father in order to gain a kingdom.

Ahmed Shah had put to flight a band of robbers who lay in Vigne's path, and the explorer took the opportunity to sketch the imposing leader on the field of battle. On the three-day march across the desolate Deosai plateau following their encounter, an enduring friendship was established between the two men. The Baltis are a hardy race, living off an unchanging diet of atta and apricots, far less food than other tribes in these regions. Defiant, strong and constantly waging their long-drawn-out battle against nature, they are considered by many to be even better mountaineers than their Nepalese equivalents, the Sherpas. As Vigne reached the edge of the plateau and looked down into Baltistan, he saw the last and most imposing of all ranges barring the route to the north, the Karakoram. From the fort of Skardu he attempted to find a chink in the world's greatest natural armoury, combining in its defences snow, rock and ice precipices, high altitudes, long distances and several less foreseeable hazards including mud flashes, swollen rivers and rock avalanches. Skardu lies on the south side of the Indus, opposite the Shigar valley to the north, which forms the route to K2, Gasherbrum, Chogolisa, the Muztagh Tower, Broad Peak and a host of other mighty peaks. To the east lie the narrow tracks to Saltoro Kangri, Masherbrum, K12, Teram Kangri and the peaks of the Siachen Valley. To the west are Gilgit and the peaks of Nanga Parbat, Rakaposhi, Haramosh and many more, whilst still further to the west is Tirich Mir.

On one of his forays out of Skardu, Vigne attempted to reach Gilgit, but as he approached Bunji from Astor, several days' march out of Skardu, he found his expedition brought to a halt by the unfriendly relations between the Gilgitis and the Baltis. The crossing of the Indus at this point proved to be a major obstacle in future explorations of the area. On his next journey he attempted to reach the Hispar valley and thence Hunza via Shigar, but although well acclimatised, fit and determined, he was defeated by the glaciers and climate of the area.

In his next attempt he travelled upstream along the banks of the Indus and thence up the

Shyok, hoping to discover the Ladakh-to-Yarkand (China) route so persistently sought after by Moorcroft. However, his previous successful explorations had now been reported to, and queried by, Ranjit Singh and he found himself temporarily 'delayed'. By the time he was released from this period of 'protection' against the Ladakhis, the few weeks of summer had passed and he could only travel as far as the Nubra Valley, into which the mighty Siachen glacier pours its river. Furtively, he attempted to reach the Siachen Valley again via the village of Khapalu, the Shyok river and the Saltoro Valley, but once more he was defeated by the weather.

In spite of this, Vigne's exploits meant that a European was now able to give an account of the southern ramparts of the Karakoram and to fill in some blanks, thus exciting the imagination of countless geographers, geologists, mountaineers and naturalists.

At this time the power of Ranjit Singh was at its zenith, but from Vigne's experience in Kashmir, Baltistan and Ladakh and from subsequent negotiations in Lahore concerning permission to travel, it was increasingly obvious that the controlling power was subtly but inexorably slipping into the hands of the Raja of the state of Jammu, Gulab Singh, who, together with his brother, controlled large portions of Ladakh and the Pir Panjal. If Baltistan succumbed, he would enclose the Vale of Kashmir from north, east and west. It was not surprising therefore that when the Maharaja of Lahore died in 1839, Gulab Singh became the first Maharaja of Kashmir. He, too, was not keen to let travellers pass into areas over which he had only nominal control, another important factor in future exploration.

Earlier, a most remarkable explorer had come to light, one to rank higher than Moorcroft or Vigne if his story be true. Alexander Gardiner apparently had several treks to his credit in the Western Himalaya and had visited Gilgit ten years before Vigne's unsuccessful attempt to reach that village. Unfortunately much of his story is unsupported, but today there are sufficient threads of evidence to give not only credence but authenticity to his travels. Certainly he deserves a meticulous study by some enterprising PhD student prepared to combine literary skills with detective work and a not insubstantial experience of living and travelling in high moutain ranges.

The doubts surrounding this most enigmatic of all travellers across the Hindu Kush, Pamirs, Karakoram and the Pir Panjal stem as much from the nature of the man as from the nature of the country. While most if not all explorers come from without, Gardiner came from within. While most returned home to tell their stories to their family and friends, he lost his family in the mountains and never returned to Europe to report his extraordinary tale. His travel notes are sketchy in the extreme and accounts of whole days and sometimes weeks are void of detail. To the avid public, desirous of reading all they could about the territory, this was an almost inexcusable lapse and, when compared with the extensive notes of Moorcroft, was considered highly suspicious. Be that as it may, while explorers were trying to break the riddle of the Karakoram by travelling south to north, Gardiner came from the point of inaccessibility in the north, Yarkand, to arrive in Kashmir in the south.

Most explorers know they will be asked many questions when they return home, and make a detailed record of their travels; Gardiner, however, lived in these mountains; they were his home; and perhaps, like many a seasoned mountaineer, he did not feel required to record or to expound on his own personal feelings or daily movements. The vast majority of people living in more civilised and sophisticated surroundings do not make daily notes, simply because their lifestyle appears to them to be commonplace. So it was with Gardiner.

Born in America in 1785, he had travelled in the north and south of that country, visited Spain, Ireland, Egypt and the Black Sea and then, at the age of thirty, arrived on the Russian shores of the Caspian Sea. After assisting his brother in the Russian town of Astrakhan, he was forced to leave after his brother's death and spent the next thirteen years journeying south to Herat and north to Khiva and the Aral Sea and visiting many of the tribes to the west of the Hindu Kush. Later he visited Ura-Tyube near Samarkand in Western Turkestan. Bearing in mind his restless, nomadic life, during which his journeys totalled many thousands of miles, it is hardly surprising that he and his followers were considered outlaws in their day. Travelling south, he crossed the Oxus at Hazrat Imam, shot three men from the ranks of Murad Beg's band and was eventually arrested by the leader of a group of Afghan rebels, Habid Ullah, who at that time was conducting yet another guerrilla war against the ruler of Kabul, Dost Mohammed. Joining the rebel ranks, he took as his wife a captured princess who bore him a son. Undoubtedly this was the happiest period of his buccaneering life, but tragedy came when both his wife and son were horribly slain while he was temporarily away defending his chief. Wounded, he and his own band of followers now had to flee, with nothing but their skills as brigands to rely on. Waylaid on their way north to the Oxus, their numbers were reduced from thirteen to only seven as they battled through one of the Hindu Kush passes. Eventually they reached Wakhan, crossed the Oxus near the Shahkdara and headed on for the supposed safety of the Pamirs. Following an unknown route they eventually reached Yarkand. Unfortunately no record of this incredible journey was kept and much has to be taken on trust. The same applies to Gardiner's seven-week journey south from Yarkand to Leh over the Karakoram and down to Srinagar. To stretch the imagination to the limit, he then returned to Afghanistan via Gilgit, Chitral and Kafiristan – a most remarkable achievement as far back as the 1820s. It was to be another twenty years before Thompson crossed the Karakoram pass from the south; forty years before Shaw and Hayward entered Yarkand, and fifty years before another Englishman could report on those most isolated of tribes, the peoples of Kafiristan.

Into the field now came Lieutenant Alexander Burnes of the Bombay army. His book *Travels into Bokhara* put the whole area into perspective for the first time, although in fact he never set foot in the high mountains. With his skills as a scholar, diplomat, soldier and author, Burnes could tell exactly the kind of story people were avid to hear; unlike Gardiner, he was reporting on strange and most unlikely events, incredible landscapes and a political scene that still requires the imagination and skill of a detective to unravel. Furthermore, he was blessed with an energetic surveyor, Dr Gerard, who produced a map measuring $3m$ by $1m$ ($10ft$ by $3ft$), to a scale of approximately $1:300,000$, that linked India to Central Asia. He made people aware of the Leh (India)-Yarkand route and the incredible hardships involved in such journeys, and also focused attention on Yarkand and Lake Sir-i-Kol. The former was the fleshpot of Eastern Turkestan, an open city and the terminus of the trade routes from the north, south, east and west, whilst the latter, situated on the windswept barren plateau of the roof of the world, was reputed to be the source of the Oxus, the Jaxartes (or Syr-Dar'Ya) and the Indus. Here was a mystery that provoked many questions and aroused more speculation than that surrounding the exploration of the Nile. Could a single lake have three effluents, each of which eventually became a major river of Asia? Where did the supply of water required for such a source come from? What was the obviously special nature of the area that could produce such a unique geographical feature? Burnes also focused attention on the problems of demarcating international borders and the requirements for defining zones of interest. He started the 'Gilgit Game'[1], the process of political, military and economic manoeuvring by the powers concerned to gain control over still undefined territories.

At this time British India did not include Kashmir, Ladakh, Baltistan or the other provinces of what is now Pakistan. On the other side of the mountains, Russian territory did not include Eastern Turkestan, the Aral Sea, the Caspian, the Kazakh steppes nor all the tribal areas to the south, but it was believed then, as now, that the Indian subcontinent could be invaded by the Russians via Kabul. What made matters worse in British eyes was the relative ease of advance from the north in comparison with the long and difficult lines of supply and communication available to the British army. When one is in a boxing ring of unknown dimensions, blindfolded and none too sure of the assistance of seconds, the difficulties of one's supposed 'enemy' are given scant attention: survival is probably the only thought in one's head. Burnes' task was to determine the size and shape of the ring, seek the allegiance of any of the seconds and remove the blindfold.

His job was difficult. He had to tread warily between the enmities of Sikhs and Afghans, to report on a threatened liaison between Russians and Persians and to negotiate with his friend Dost Mohammed in order to bring Kabul and Kandahar into the British zone of influence. Fortunately he was also able to have friendly discussions with the old enemy Murad Beg who, in his old age, was anxious that the expedition doctor, Percival Lord, treat his ailing brother, who most unusually had remained a life-long ally. Lord was accompanied by Lieutenant Wood, a naval officer, who had the unlikely role of finding the source of the Oxus within the highest land mass on earth. Wood's journey, undertaken in winter, was excruciatingly difficult. Temperatures of $-14°C$ (6°F), frequent blizzards and the effects of altitude all hindered progress. Many dangers were reported by him, but the one most feared by any traveller in these valleys is a boulder avalanche down the high and steep valley walls. Shortly before, one party had lost half their number, plus mules and baggage, all of which were swept into the roaring torrents of the river.

At Qala Panja, the confluence of two great rivers, Wood had a momentous decision to take: which branch was the Oxus? After taking some measurements, he decided to take the northerly route to Sir-i-Kol Lake and the Pamirs, rather than the Sarhad river north of the Hindu Kush. Helped by the nomadic Kirghiz, he met and overcame further hazards before reaching the lake on 19 February 1838. There was only one outlet, a mere trickle in the frozen winter months. The lake was surrounded by extensive and uninterrupted snow fields.

While this discovery was being recorded, Burnes had been recalled. Although he had never had explicit instructions, by his actions he had become increasingly involved in the political manoeuvres of his own government. He had promised to help his friend Dost Mohammed to claim Peshawar and when this plan was rejected by his superiors, he then did an about-face and assisted the Indian Army into Kabul, ousting Dost Mohammed and placing Shah Shuja, a most ineffectual man, on the throne. Two years later Dost Mohammed and his son, continuing the struggle from the foothills of the Hindu Kush, massacred the officers of a British patrol that included Lord. Several months later, soon after the route to the Khyber Pass was blocked by the tribes around Jalalabad, Burnes himself was killed. Dost Mohammed was reinstated, the British suffered their greatest defeat, and Afghanistan was now closed to all travellers. It was not until 1894 that Lord Curzon, after much research, pronounced that the actual source of the Oxus was the Wakhjir Glacier, north of Hunza. From this glacier emerges the river that flows to Qala Panja, there joining the river from Sir-i-Kol Lake.

In 1841 a momentous event took place; forces of Gulab Singh, Maharaja of Kashmir, invaded Tibet under the leadership of Zorawar Singh. Gulab Singh had now gained control of Ladakh and Baltistan and marched up the Indus from Leh with six thousand men. In his ranks was Vigne's old friend, Ahmed Shah, Raja of Baltistan. Gulab Singh's aim in attempting to annex Tibet into his growing empire was to control the wool trade, used for making shawls in Kashmir, and to cut off that portion of the supply from Tibet which went to the British dependency of Rampur, east of Simla. The prospect of plundering the Buddhist monasteries was added incentive for the Hindu forces, now more commonly known as Dogras. At that time the British were trying to conduct a treaty with the Chinese and so protested strongly, conveniently forgetting their own Afghanistan campaign.

The invading southern forces knew little of the military tactics necessary for the conquest of Tibet, nor presumably of the terrible climatic conditions, and were repelled by Chinese forces, who descended as far down as Leh to take control of that city until the Dogras returned in force and signed a treaty. This involved the British taking responsibility for defending the northern borders and arbitrating in disputes. An official expedition was mounted to enable the British to familiarise themselves with the Ladakh-Tibet borders and also, it was hoped, to bring China more directly into touch with British interests. Unfortunately the British reckoned without the intransigence of the Tibetans, who are always suspicious of strangers and who, even after the recent war, had no incentive to change the well-established traditions of trade between the two Buddhist centres of Leh and Yarkand.

The British boundary commission failed in its major task, but succeeded in two important projects. Alexander Cunningham, leader of the project, brought back a most comprehensive report on Ladakh, while Dr Thomas Thompson, a distinguished naturalist, wintered in Baltistan and attempted to reach Gilgit. The local war between the Dogras and the Dards, the latter being a collective name for the peoples inhabiting areas around Chilas, Gilgit, Yasin, Hunza and Chitral, obliged him to return to Skardu, but undiscouraged, he applied for permission to continue and eventually reached Leh, from there setting out to cross the mountains to the north. Thompson had little awareness of what lay in store, other than the false, indeed potentially disastrous view that the crossing of the Karakoram Pass would permit him entry into Central Asia. From Leh, he surmounted the Khardung Pass at 5,330m (17,500ft) and then, leaving the Nubra Valley, the Sasser Pass, which is reputed to be the most dangerous of all. Here he found bones and skeletons littering the track and acting as route markers. He was now on new ground, prospecting the route sought by Moorcroft and Vigne. The prospect was bleak in the extreme and confirmed the unsupported views expressed by Vigne, who explored north and east of Skardu, that here was the ultimate barrier – a barrier formed of glaciers, unstable moraines and the highest plateaux in the world, and suffering the worst possible weather. This stark, daunting, grassless terrain required of the explorer both a determination to meet the physical challenge, and the faith that somewhere this harsh landscape must yield to something more hospitable, where life could be sustained. For here there was evidence in abundance to show that many had perished, in some cases from the effects of altitude alone. It is little wonder that having reached the watershed between the two continents, Thompson, suffering acutely from headaches, decided to turn back.

From the 5,550m (18,200ft) high Karakoram Pass, Thompson saw none of the things he had hoped to see: the cities of Central Asia, the passes that lay ahead or the route to those remaining passes across the continuing reaches of the highest plateau on earth. Perhaps the most impressive fact to be learned about these mountainous tracts, even to this day, is that there is no single pass to the other side, but an almost unending series of peaks and cols. Any mountaineer who comes to explore the Karakoram must accept the fact that great efforts on his part may be rewarded by an apparent lack of progress. But Thompson, like many to come after him, had done enough to tire the body, excite the mind, and receive the acclaim and adulation of the awaiting public, who by now were becoming more aware of the vastness of this highest of all mountain ranges, the Karakoram.

Eventually, in 1857, Yarkand and Kashgar were reached, but by a German, Adolph Schlagintweit. In the previous year, his two brothers had breached the Kun Lun passes and returned safely without blundering into the civil war that probably precipitated the assassination of their brother the following season. Their efforts were recognised in 1859, when the Tsar of Russia conferred the title of Lord of the Kun Lun on Robert Schlagintweit. The implications of such an award did not go unnoticed. And indeed, at about this time the British were intent on gaining a more accurate assessment of the terrain.

Of interest to the 1980 project was the fact that in 1855 the British Grand Trigonometrical Survey, GTS, commenced its most difficult and most ambitious project yet, a survey of the northern areas of India. Started in 1800 in Madras, it was to culminate with the work of Mason in 1913, which linked the Indian and Russian systems so as to provide the most extensive land survey line anywhere on earth. Crossing the Pir Panjal was incredibly hazardous. Nevertheless, before 1860, both K1 (Masherbrum) and K2 were known to be 7,802m (25,600ft) and 8,611m (28,250ft) respectively. It was now clear that here was the world's highest mountain range.

'K2' has remained that mountain's official designation, and this is the name used by both mountaineers and the local Baltis. Attempts to call it by any other less abstract name are regarded by most as an affront to the skills of all those explorers, surveyors, mountaineers and scientists who shared and will continue to share in the exploration of the Karakoram and stare with awe at this world's most singular, pyramoidal and dangerous peak, even standing as it does amongst many other giants of similar proportions (see map, p. 6). The name of K2 is slightly mysterious and frightening. It demands the attention it deserves, and no man, even though he treads its summit snows, will ever conquer it or know it in all its dangerous moods.

Colonel Godwin-Austen was the first surveyor to determine the position and height of K2 and seek the opportunity to cross the watershed. During his first season he was beaten by angry villagers, a not unusual hazard of the area, but after a temporary retirement from the service he was back, this time to witness a stream flash caused by the breaking of a glacier dam, in which a black, heaving mass of mud containing rocks many metres in diameter poured down past his campsite, close to a ravine. The mud flow was some 30m (100ft) wide and more than 1½m (5ft) deep. Such flashes are frequent and add yet another evil threat to the many others that cause death and destruction in the villages of the Karakoram.

After braving moraines, and crevasses of unknown depth that punctuated a difficult ascent of the Panmah Glacier, Godwin-Austen just failed to reach the top of the Muztagh Pass and with a storm at his heels thankfully regained his campsite. It would not be until 1887 that the pass would be scaled by Sir Francis Younghusband, and the 'trade-route' to Yarkand be dismissed as the regular route in undoubtedly had once been, albeit on a very limited scale, because the advance of the ice had made the route impractical.

One extraordinary man of this period was William Henry Johnson, who was both an experienced surveyor and a mountaineer. Not only did he attain a height of 7,010m (23,000ft) in the course of his work, but he also travelled to Khotan when the triangulation work of the GTS was coming to a close. He did not quite reach Yarkand, but wisely returned to Ladakh in late 1865, via the Sanju and Suget passes of the Kun Lun. Johnson reported that

Russian caravans were now regularly penetrating as far as Khotan. As with the pleas of Moorcroft, the British government paid little attention to Johnson's suggestions that they extend their influence into Eastern Turkestan (now part of Sinkiang), where Chinese control was on the wane. When the government refused to support him, he retired into the Maharaja of Kashmir's service. No other person in the service of the GTS had been more frequently overlooked or so obviously succeeded in promotion by men of inferior experience but higher social position. Undoubtedly Johnson was not helped by the fact that he was not from the ranks of the ruling military classes.

In 1868 Yarkand and Kashgar were finally reached from the south by two British explorers, Shaw and Hayward, who, although they had much to gain by collaboration, chose to go their separate ways along the same initial route – that used by Johnson, east of the Karakoram Pass. Although avoiding the deep, uncertain paths of the gorges of the Shyok river, this route climbed much higher passes, namely the Chang-La, the Marsimik and the Changlung, before crossing almost 320*km* (200 miles) of desolate plateaux, the Lingzi Thang and Aksai Chin deserts. Shaw, the more commercially-minded of the duo, was 'delayed' before the Sanju Pass until Hayward arrived, but was then given permission to proceed. Hayward, too, was 'delayed' but, being the more aggressive of the two, he managed to slip away and travelling west in November–December at an incredible speed for this terrain, altitude and time of year, traversed 480*km* (300 miles) in twenty days. After being hunted by soldiers, with only brigands crossing his path, and being forced to eat his pack animal, he found the source of the Yarkand river. Although the soldiers eventually caught up with him, he had by then obtained permission to proceed to Yarkand. In his travels, he had reconnoitred new passes in the Karakoram and Kun Lun and so had proved to himself that he had the necessary exploratory skills. He now pressed on without further delay towards his major goal, the Pamirs.

In Yarkand, then under the control of Yakub Beg, ruler of Eastern Turkestan, both Hayward and Shaw lodged a while but they never consulted each other. They both wanted the exclusive explorer's prize. However, they were now in dangerous country, under strict Mohammedan rule; few travellers escaped from it and treachery and assassination were commonplace, the resident Chinese having been speedily despatched by murder or starvation. Yakub Beg was a harsh interpreter of Muslim law: mutilation and flogging were common events; women were put in purdah, and prostitution banned – although harems still flourished.

On reaching Kashgar both Hayward and Shaw were imprisoned. This period of imprisonment was difficult for a man of Hayward's temperament, since the Pamirs could be seen only 100*km* (60 miles) away, but Shaw only wanted to return to India and his spirit flagged; eventually he became compliant to the wishes of his captors, no doubt thinking this a more profitable stance than the angry outbursts of Hayward. Both men were surprised when they were released – an event that was probably precipitated by the advance of the Russian empire, which had now reached Tashkent and Samarkand. Eastern Turkestan was next in line, and Yakub Beg needed allies; as a result, the status of his two English prisoners suddenly changed. Britain was now deeply concerned at the possibility of the Russians invading the Punjab with possible assistance from Persians, Turkomans and Cossacks, so in order to divert Russian threats from this area, attempts were made to establish commercial links with Yarkand via the Karakoram Pass.

When he was released, Hayward took the opportunity to visit the vast and as yet unknown area in the north-west of the Karakoram, where the range joins the Hindu Kush and Pamirs.

This area would be the last to yield its geographical secrets. More remote, more difficult of access than any other area in what was now known to be the most inaccessible zone on earth, the land of the Dards would be defended from within and without with a frightening intensity. It was this territory that became the site of the 1980 Karakoram Project.

The Maharaja of Kashmir, as jealous and suspicious as ever and keen to limit the activities of Shaw and Hayward, knew that both could, and probably would, weaken his authority in Kashmir. They could also expose his weaknesses in the surrounding territories to the west and north, and compete with his own commercial interests. He planned, therefore, to attack Hayward, who had further angered him by revealing a mass murder perpetrated by the Kashmiris in Yasin, and to abandon Shaw, who was assisting a new British initiative to Yarkand, led by Douglas Forsyth, a District Commissioner in India who had long desired an opportunity to lead a British team to Eastern Turkestan. Shaw and his companions almost starved to death on the Karakoram-to-Kun Lun plateaux when the scheduled supplies were not forwarded. Hayward was to fare much worse several hundred kilometres to the west.

During the Forsyth mission to Eastern Turkestan in 1873, one party was permitted to journey north into the Tian Shan. They reached the Russian border, and in so doing, joined up the survey work of both nations. At last the two empires had met.

In 1874, Lieutenant Colonel (later General) Sir L. E. Gordon led a small party into the Little Pamirs, via Tashkurgan. Their aim was to investigate the questions posed by Wood's journey thirty-six years previously concerning the hydrography of the area, and to elucidate the nature of the ground travelled by Gardiner. They found another source of the Oxus, the Aksu, but this was somewhat distressing since it meant that the agreement London and St Petersburg had just signed, defining the Afghan border as the Oxus river, was now meaningless. Worse still, although they found the going tough in early spring, it was obvious that in mid-summer there were a variety of passes through the Pamirs that were all much simpler and less hazardous than anything in the Karakoram. It appeared that this was the easiest route for Russian expansion. Captain John Biddulph, a member of the expedition, knew that the passes to the south from this region led quickly and easily to Chitral and Gilgit. Meanwhile Gordon himself examined a more direct route to Gilgit, which lay down a valley called Hunza.

In 1877, the despotic ruler of Eastern Turkestan, Yakub Beg, was murdered and the Chinese returned to rule, renaming the state Sinkiang, meaning 'The New Dominion'. Understandably, they met no opposition from the local population. However, because of the recent links forged by the Forsyth mission, the Chinese terminated all contact with the south.

Thus the fate of the Karakoram Pass as a viable commercial route was now sealed, especially as the one man formally most responsible for keeping the route open for trade, Andrew Dalgleish, was murdered on the pass itself by an Afghan. The murderer was eventually located in Samarkand, where the Russians, who now had a working relationship with the Chinese, reported that he had taken his own life.

## Gilgit and Hunza

The final act in the exploration of the Karakoram began in 1866. Soon the route through to the very heart of the Karakoram would be known to any who cared to risk their lives following it. Due to the changeable whims of neighbouring governments, however, policies would fluctuate between restricting the movement of their own nationals (other than special investigators), and despatching armies to control what they thought to be strategic and commercial centres.

Although Gilgit and Hunza had not been visited by Vigne, and Gardiner's route remained unknown, the long route across the desolate Deosai Plain to the Indus at Bunji, south of Gilgit, was relatively straightforward in the summer months. Beyond lay unknown territory. Gilgit is situated close to the entrance to the Hunza Valley, but because of the nature of the terrain it is hidden to all eyes until the last 5km (3 miles) of the approach track. Its geographical position can only be described as breathtaking. To the south of Gilgit, stretching east to west, is the start of the Hindu Raj, a mountain range that blocks off Chitral to the west from the Indus and Swat valleys, while further south lies Nanga Parbat, the terrifying remnant of the Great Himalaya. The confusion of high peaks to the north marks the meeting point of the Hindu Kush and the Karakoram. To the east, the only route lies up the Indus to Baltistan, Ladakh and Tibet, whilst to the west are higher passes that lead to Chitral and Afghanistan. To approach Gilgit by following the Indus river up from the Punjab plains would have been impractical, since the river gorge necessitated treks up and over countless ridges and in and out of innumerable ravines. A full day's effort would only have yielded a few kilometres' progress upstream to a bivouac site that could provide no shelter or succour. Little wonder that Gilgit and Hunza contained one of the last pockets of humanity to survive the ravages of 'farangi rok', the foreigner's venereal disease. This was the area into which Gottlieb Wilhelm Leitner, a resident of Lahore, stumbled during the summer of 1866.

Leitner came down to the village of Bunji on the Indus from Kashmir, by following the Astor river downstream. His aim was to study the languages of the peoples of the area, whom he was the first to classify as Dards – presumably because he thought them to be descendants of the Daradas of Sanskrit literature and the Daradae of classical geographers. Officially, he was there to examine the possibility of Chilas being 'Kailas', abode of the Gods of Hindu mythology. Leitner was exceedingly well qualified for the task, since he spoke twenty languages; he was also probably the first visitor to be committed to conservation, and was ridiculed for being so. His attempts to understand, document and preserve local and national cultures were partially sabotaged by the Maharaja of Kashmir, Gulab Singh, who did not want his precarious hold on the territory and its many tribes, some of whom were unknown to Europeans, to be investigated. There were, for example, reports dating back to before the days of Alexander the Great of a blue-eyed, fair-skinned people with blonde or reddish hair, isolated for centuries in a vast, mountainous eyrie.

Leitner was horrified to learn of the atrocities committed by the warring factions who were so ethnologically estranged: captured Dogra troops of the Maharaja of Kashmir were burnt alive by the Dards, and when Gilgitis and Hunzakuts were taken prisoner they were sold as slaves. When Leitner entered Gilgit, now garrisoned by Kashmiris (Dogras), he found it deserted, but he eventually located a few Dards and thereafter they became his adopted people, even though he spent only thirty-six hours in the area before wisely retreating back to Lahore. In this short period he was made aware of the highly volatile, politically important zone which he named Dardistan, and which included Chilas, Gilgit, Yasin, Hunza and Chitral. Both Kashmir and Dardistan were at present beyond the area controlled by the British government, but they could no longer be ignored, since it was soon to be obvious that some line needed to be drawn on a map that would clearly separate the Chinese, Russian, Afghan and British zones of influence. Unfortunately, even the local Afghans were unaware of the territories which were subject to their own rule, and another problem was the uncertainty surrounding the Oxus river and its true source.

In 1869 and 1870, George Hayward continued his explorations by trying to enter the Pamirs from the south. Before travelling to Yasin he waited for local news from the valley in

Gilgit; he knew the human barriers that lay before him in terms of the different languages, cultures and religions of the warring Dogras and Dards, for whom perfidy and atrocities were the order of the day, and knew also that neither side would trust him. But he now had the additional problem of waiting until the winter snows cleared. The ruler of Yasin, Mir Wali, was anxious to control Gilgit, especially since the Dogras, as the Kashmiris were called, had afflicted appalling atrocities on the beleaguered Yasins, some 1,400 of whom were reported by Hayward to have been massacred. Though totally aware of the dangers of the territory through which he had to travel, Hayward, unlike Leitner decided to go on. He was murdered in Darkot on the morning of 18 July 1870. He had sat up all night in his tent with a rifle and pistol nearby, but in a drowsy moment in the early morning sun he was overpowered and killed.

Hayward's murder still raises many questions. Why did the supposed inciter, Mir Wali, kill the man whom he had once taken out as a friend on hunting expeditions? One official report gave four good reasons why Mir Wali should be suspected, but common sense would discount four reasons; one should be enough. Most suspicion was eventually directed towards Mulk Aman, one of Mir Wali's brothers ousted from Yasin, who was now in the service of the Maharaja of Kashmir in Gilgit. Certainly the Maharaja did not wish the British to become aware of his misdemeanours. Whatever the truth of the matter, the grave in Gilgit cemetery is a permanent reminder that although most men try to seek friendship and companionship, personal ambition and misunderstanding can be as deadly as any of the other dangers that abound in these mountains.

The next Englishman on the scene was John Biddulph in 1874, who was entrusted by the Indian government with the mission of determining how and why Chitral and Badakshan, and Hunza and Sinkiang had close ties with each other across the Hindu Kush and Karakoram watersheds respectively. Biddulph was the first spy to enter this arena. Previously, Colonel Gordon's party (of which Biddulph had been a member) had indicated the ease with which the Russians could attain the northern slopes of the Hindu Kush, particularly in comparison with the British, who would have great difficulty in reaching the southern slopes to defend what were now known to be relatively simple passes into the subcontinent. The agreement on zones of influence accepted by Russia and Britain was now meaningless, firstly because the Afghans had far more control of the tribes on both banks of the Oxus than had been hitherto appreciated; and secondly, because the river had been incorrectly located on the maps on which the agreement was based. As Russian influence advanced south and east, Sinkiang was now being threatened, and soon it would be India that would be under pressure. It therefore became necessary to have detailed geographical, political and economic information about Hunza.

Knowing the fate of Hayward, Biddulph did not advance up the valley until a hostage was safely housed in Gilgit to ensure his own safety. The Mir of Hunza, Ghazan Khan, was custodian of the route through Hunza to the Kilik, Khunjerab and Mintaka passes, but at that time the three large villages at the lower entrance to the Hunza Gorge, namely Nomal, Chaprot and Chalt were not under his control (see page 100). The latter two villages were aligned with his neighbours and bitter enemies, the Nagaris, whilst the former was dominated by the Gilgitis. Ghazan Khan wanted Biddulph to use his offices to secure a transfer of authority, but Biddulph refused and was virtually made a prisoner. Eventually, however, his implacable and obstinate stance succeeded and he was allowed to return to Gilgit. Naturally it was Biddulph's opinion that the sooner all the gorges leading up to the frontiers were guarded by troops loyal to the Indian government (and this meant by

Kashmiris) the better, but that this control should be tempered by the presence of a British political agent in Gilgit. The importance of Gilgit as the nerve centre of the Karakoram was now established.

From Biddulph's experiences, the geographical setting of Hunza was also at last recorded, but who would believe the reports? Who could write descriptions of scenes that defy description? Authors can only use words and photographs, only present two dimensions. Eric Shipton saw the Hunza valley as the 'ultimate manifestation of mountain grandeur – the most spectacular country I have ever seen'. More recently, John Keay, author of the delightful book *The Gilgit Game*, was equally overwhelmed: 'There are no pools for the fisherman, no falls for the photographer and no grassy banks for the picnicker, just this thundering discharge of mud and rock'. The Hon. George Nathaniel Curzon wrote the following: 'The little state of Hunza alone is said to contain more peaks of over 20,000*ft* [6,100*m*] than there are over 10,000*ft* [3,050*m*] in the entire Alps. The longest glaciers on the globe outside of the Arctic circle pour their frozen cataracts down the riven and tortured hollows of the mountains. Avalanches of snow, and still more remarkable of mud, come plunging down the long slopes and distort the face of nature as though by some lamentable disease.' He continues to describe 'this great workshop of primaeval forces' by remarking that here 'nature seems to exert every note in her vast and majestic diapason of sound.'

All these descriptions, whilst true, are inadequate. Hunza is only one – albeit the most important – of a network of gorges down which rivers of liquid mud flow. The Shimshal and Hispar valleys are equally impressive and disgorge their angry floods into the ravine that divides the Hindu Kush from the Karakoram. They also have their own network of chasms, over which walls of rock, snow and ice tower as high as 6,000*m* (20,000*ft*) above, precipitating avalanches of débris that plunge apparently noiselessly across frail tracks to disappear into the dark, brown depths of the cold, unsympathetic, turbid flow.

The word 'track' in this environment is somewhat misleading, even to those who have beaten out furrows in drifting snow over melting glaciers. Here in Hunza, a track is frequently the only line of advance, and when it is destroyed a totally different approach needs to be explored, frequently over distances of a few kilometres. Sometimes a track will disappear into scree several thousand metres high, continually shifting downslope. Sometimes the track will be no more than a series of shaky boulders down at river-level before rising to climb near-vertical dry mud walls several hundred metres high. To follow such a track is nerve-racking in the extreme, and more than one rock-climbing athlete has prayed for the comparative safety of a vertical rock climb on the cliffs of North Wales or the English Lake District. When a track crosses vertical rock walls, boulders wedged in cracks and mud cement are utilised to provide a rickety stairway. High above the valley floor the river may no longer be heard, but it remains omnipresent, reminding one of the penalty for a single false step.

The village oases in the Hunza valley give transient shelter both from the dangers of the river below, which erodes away the edges of cultivated land, and from the cliffs above, down which falls an endless stream of rubble from the disintegrating walls. Here one can rest and recover from the shocks of the journey before pressing on to the next stage. Little wonder that Major General Shahid Hamid, Pakistan Minister of Information, considered Hunza to be as well-protected as Shangri-La.

Having once accepted Gilgit and Hunza as the cornerstone of a defensive position, it became necessary for the British forces to define the defensible border; and herein lay the major difficulty: none of the fiercely independent tribes would give perpetual allegiance to any major state. The Amir of Kabul wished to have Badakshan as an ally, while the leader of Badakshan wanted to control the states of the upper Oxus. The tribes of Afghanistan also wanted to absorb Chitral, whose ruler, the Aman-ul-Mulk, had for a long time had influence over Yasin. This unstable situation effectively eroded the defensive line of the Hindu Kush. In truth, all parties wanted to control; none to *be* controlled. The same state of affairs remains to this day.

Biddulph was the first political agent in Gilgit, and soon realised his mistake in trusting the Maharaja of Kashmir. His suspicion was vindicated in 1879 when the British eventually invaded Kabul (where a Russian mission had been installed) and correspondence was found there showing the duplicity of the Kashmiris. The Afghanistan offensive had also been precipitated by Russian manoeuvres across the Alai and Pamirs towards the Hindu Kush passes, although these had failed because the attempts had been made in May, just a few weeks too early. The British offensive itself was none too successful and became bogged down in the complexities of Afghanistan affairs.

Back in Gilgit, the Yasin tribes advanced down the valley at the same time as the Mir of Hunza threatened Chaprot. Unexpectedly but fortunately, the Chitralis came to the rescue and deposed the Yasin ruler. Biddulph could now breathe again and attempt to take stock of who was enemy and who was friend – at least for the next few weeks. The cultural, geographical and political interactions in the area were at last being appreciated, if not quite understood.

The next move in the Gilgit Game was to survey Kafiristan, a state bordering Chitral, completely unknown to Europeans. Of all the ethnic puzzles of the Hindu Kush, this was the greatest of them all.[2] The religion of the Kafirs was pagan, their language unknown, their allies non-existent. Certainly they were not Dards. More information was needed, and so, under the pretext of sabbatical leave from the survey of India and in the guise of a doctor of medicine, William Watts MacNair, alias Mir Mohammed, travelled with his tools of trade to unlock the secrets beyond Chitral. Crossing the Malakand Pass into the Swat Valley proved easy, the Lowarai far more difficult, but in Chitral he was well received by the Aman-ul-Mulk. Exploration to the north-west showed that the Dora Pass led with ease into Badakshan, but that the Shawal Pass into Kafiristan (now called Nuristan) was more difficult. Sadly he was only permitted to stay a few days.

As British governments changed, so did Indian policies, and so in 1885 Colonel William Lockhart was sent into Dardistan with a military mission led by four British officers, to find answers to several intriguing questions that MacNair had introduced into the equation. It was also certainly Lockhart's mission to find the means of bringing Chitral into the British zone of influence and thereby thwarting Russian ambitions. Unfortunately, no sooner had the party moved out of Gilgit than catastrophe struck: a mule carrying 4,000 rupees in silver fell down a cliff face into the river. Such a disastrous start mirrored Leitner's experience of losing his companion, Henry Cowie – Cowie having insisted on crossing the Dras river by a rickety bridge over a gorge on the back of an equally unfortunate donkey.

The Lockhart mission studied many of the passes leading to Russia and proposed that the Aman-ul-Mulk should provide the first line of defence against possible Russian advances. As this Chitrali leader was now estimated to be about seventy years of age, it was important that Lockhart should identify which of his sixty sons would be the future ruler. Lockhart unfortunately backed the wrong candidate. Similar negotiations to those concluded in Chitral were conducted in Kafiristan, but it was obvious that defence would be difficult in view of the hostility between the tribes. At one point the Kafiris took one of the leading

Chitralis and threw him over a cliff, telling Lockhart that should he return, he must come without the Chitralis.

Onto the exploration scene now burst Ney Elias, a British traveller of vast experience who, in the same period as the Lockhart forays, crossed in one circuitous journey the Pir Panjal, Great Himalaya, Karakoram, Kun Lun, Pamirs and the Hindu Kush. Small, flexible expeditions whose members could move fast were the secret of his success. In Sinkiang, the Russian Consul in Kashgar was eager to despatch this single European, who he feared could create untold difficulties if not constrained. Nevertheless the Chinese permitted him to leave by way of the Pamirs, and within ten weeks he became the first man to cross the area from east to west and to recognise all the northern approaches to the line of Indian defence.

By this time, however, Elias was a dying man, and he had a terrible struggle to reach and report to the British Boundary Commission in Afghanistan, which was attempting to define the borders of India and its neighbours. Perhaps Elias's greatest contribution was to point out that Afghanistan could be extended to the Chinese border so as to form a buffer zone between Russia and Dardistan. Unfortunately his proposal was soon eroded down to the Wakhan pan-handle, and today that, too, has been submerged in the Russian southerly advance.

Lockhart now took the opportunity of visiting Hunza, but fell into the same trap as Biddulph. Furthermore, the Chinese now wished to keep the British out of the territory, which they considered to be under their influence. Under these circumstances Lockhart shamefully agreed to dispossess the Mir of Nagar, who controlled the two forts at Chaprot and Chalt, to the benefit of Hunza and thereby indicated that the British were willing to compromise. Soon his action would cause much suffering, and the British would have to fight hard to take control of Nagar and Hunza. Further explorations by Lockhart now angered the tribes of Wakhan. He was unable to pay his bills and enraged the Amir of Kabul, who forced him to retreat without a further visit into Kafiristan. Even the Aman-ul-Mulk hurried him back through Chitral to his post in Gilgit. Lockhart and the Boundary Commission were also rapidly decimating the meagre supplies of the valleys, thereby antagonising the local peoples whose natural sympathies went to lone travellers like Elias who required minimal sustenance.

The closing stages of this chapter in the history of Hunza were unsavoury. Elias, the one man who could have completed the essential survey of Hunza with efficiency and speed, was directly and indirectly thwarted by the rivalries of Ridgeway and Lockhart, leader of the Afghan Boundary Commission and the Gilgit Military Mission respectively, both of whom were knighted for their exploits. But meanwhile the supposed threat of invasion of India by Russia kept everyone on their toes. At last credence was given to this possibility in the person of the Russian agent Captain Grombtchevski, who was welcomed in Hunza by the new Mir, Safdar Ali, the murderer and son of Ghazan Khan. The Hunza armies were soon threatening Gilgit, while the Russians were close to annexing Sinkiang. In 1888 a turning point was reached when, by some mysterious route, the Hunzakuts restarted their plundering raids on the Ladakh-Sinkiang trade route. The Kirghiz complained bitterly to the Chinese, but receiving little support, turned to the British. Accordingly Captain Algernon Durand was ordered to Gilgit to assess the situation. The subjugation of Hunza was about to be enacted.

Two approaches were planned, one diplomatic, one exploratory. The former was to try to bribe Safdar Ali to abandon the activities which his tribesmen considered to be a traditional national sport, whilst the latter involved finding out exactly how the raiding journeys were accomplished. This would require all the skills of a mountaineer, coupled with the awareness of a military intelligence officer and the training of a diplomat. Algy Durand took the initiative for the former task, but Francis Edward Younghusband, nephew of Robert Shaw of Karakoram Pass and Yarkand fame, would be remembered longer for attempting the latter. In 1887 he crossed the Muztagh Pass and in 1889 he surveyed the Saltoro Pass. By October he had ascended the far side of the unknown Shimshal Pass and so came to discover the hidden and mysterious route in and out of Hunza. He continued his explorations to the north and north-west, and then, at a historic place called Khaian Aksai, well south of Yarkand, he, the British agent, came face to face with Colonel Grombtchevski. Here they wined and dined together and considered the future of Asia, and shortly afterwards parted as friends, continuing their explorations for their respective governments.

Despite all the evidence now available concerning the difficulties of moving vast armies through such terrain, the ambitious Algy Durand still wished to thwart the possibility of a major Russian advance through the Hindu Kush and Dardistan as a prelude to invading India. Moreover, unlike Biddulph he was now in a strong position to do something about it. He had command of 200 British-Indian troops, a mountain battery and eighteen British officers. In his view it was necessary to pacify Hunza and Nagar and stop the raids into Sinkiang. However, a subsidy of 20,000 rupees was not considered enough by the twenty-two-year-old Safdar Ali, the new Mir of Hunza, who insisted on extracting more benefits before Durand could retreat. But both knew the tribute had no binding effect, and once Younghusband had safely passed down the gorge, Durand put into operation his plan to invade a territory that was undoubtedly the greatest natural stronghold in the world. His resolve was strengthened when he learned that the raids via the hidden route were continuing and that the Kirghiz had been relieved of the money given them by Younghusband to build a better fort to defend themselves near Shahidulla.

Undoubtedly Durand was spoiling for a fight, and when Uzr Khan of Nagar murdered his own brother, a not uncommon practice, coupled with other but less serious irritations such as stopping of the mail and kidnapping, he decided to act. Further justification lay in an incident involving Younghusband, who had now been sent back to the Pamirs to continue the work of Elias in defining a border between China and Afghanistan. In this matter the British favoured the Chinese claims rather than those of the Afghans, since the former would be more difficult for the Russians to overthrow. Younghusband suggested the border should be at Somatash in the Great Pamir. Action was now precipitated by the fact that, while Younghusband was in the Pamirs, a strong force of Cossacks had taken charge of the Little Pamir to the south and now had control of the southern flanks of the Hindu Kush. Furthermore, they had crossed the Baroghil and Darkot passes and had visited Yasin. Plainly, a strong Russian military force in Dardistan could be considered a more serious incursion than the solitary Englishman to the north now attempting to drive a wedge into the expanding territory of Russia. It was Younghusband, however, who was instructed by the leader of the Cossacks to return to Chinese territory – in other words, to leave what was now claimed to be Russia. The serious implications of this eviction caused much discussion in government circles, but particularly in Gilgit. It was now thought essential that Hunza and Nagar be brought under British rule.

Durand now set about building a road from Gilgit up to and beyond Nagar and Hunza, refusing to tolerate opposition, even though he knew this meant war. The logistical problems involved in transporting provisions over from Kashmir were more impressive than the war itself; both entailed near-calamities. With 1,000 men, leaving a similar number behind to protect Gilgit from attacks from the south and west, Durand moved up the gorge to Chalt, and on December 1 crossed over to the Nagar side and up to Nilt. In spite of his superior firepower, including a machine-gun and two seven-pounder mountain guns, Durand was alarmed to find his men had little or no impact on the strong native fortifications. He was up against an experienced fighting force, who managed to keep his forces at bay by firing their motley assortment of guns through peepholes in the thick, soft mud walls of their fort. One home-made bullet, a garnet encased in lead, hit Durand in a most sensitive area and so despatched him back to Gilgit. The fort was eventually taken in a courageous action for which two VCs were later awarded, but which should also have earned a severe reprimand for the leaders of the attack.

Instead of instantaneously following up their hardwon success, the force of Gurkhas, Dogras and British waited to consolidate. The Nagaris took this opportunity to take up prepared positions immediately on the other side of the steep Nilt ravine. This position was to prove the key to the whole affair. The advance was halted for three weeks, and reconnaissances showed no way to breach the Nagaris' position, since they were surrounded by a precipice on three sides and their fortifications were continually being improved. If any evidence was required to illustrate how easily the valleys of the Karakoram could be defended, then these actions at Nilt and Hunza provided it. However, on 20 December a small force of British and Gurkhas, having climbed down the gorge to the Hunza river the previous night, now started to climb the 500m (1,650ft) precipice directly beneath the fortifications. It was a near-suicidal enterprise. Natural and artificially induced boulder avalanches, loose footholds and handholds, heavy packs of guns and ammunition all made the climb and storming of the fortifications an event that would become the central theme of many books. The individual 'sangars' (fortifications) were taken one by one. Uzr Khan of Nagar fled, and soon the British took control of Baltit Fort in Hunza, while the Mir of Hunza, Safdar Ali, fled with his women-folk to sanctuary in Sinkiang. The six-hundred-year-old kingdom was now coming to an end.

Durand continued his adventures to the south on equally flimsy pretexts, since it was as improbable that the Chilasis would take control of Bunji as it was that the Russians would advance into Hunza. However, it appeared that a direct route into the Punjab might be found south of Chilas, via the Babusar Pass. The route was shorter than the route via Kashmir, but even more important were the political implications, since direct contact with Rawalpindi would cut out the intransigent and unpredictable Kashmiris. Meanwhile, however, Durand had to recuperate in Kashmir, and the Aman-ul-Mulk, the Mehtar of Chitral, died. Afzul-ul-Mulk seized power and liquidated most of his opponenets, excepting Nizam-ul-Mulk and Amir-ul-Mulk, who escaped. But the Afzul was murdered in his turn by an uncle, Sher Afzul, previously resident in Afghanistan. Accordingly, it was thought better to place Nizam-ul-Mulk, now seeking protection in Gilgit, on the throne.

Durand first advanced on Chilas and after a bloody battle on the opposite bank, crossed the river, burned the village and built a stronghold. He then put Sher Afzul to flight and placed the unpopular Nizam in charge of Chitral. Durand now had to contend with keeping order and control in many areas of Dardistan, and the strain began to tell. Chilas revolted, and in Chitral Nizam was threatened by many enemies. At this point Durand was recalled to India; possibly a just retribution for his serious mistakes at Chilas and Chitral.

In 1894 the Hon. George Nathaniel Curzon visited the area and pronounced his solution to the problem of the source of the Oxus. He expanded Elias's correct judgement by claiming that the Sarhad was the main tributary of the Panja and the Wakhjir the main stream of the Sarhad; hence the Wakhjir is now regarded as the source of the Oxus. When Curzon left

Chitral for Gilgit and thence home, it appeared that all was well. But in January 1895, Amir-ul-Mulk became Mehtar after murdering Nizam-ul-Mulk. This in turn caused Umra Khan of Jandol, a Pathan and uncle of Nizam-ul-Mulk, to dispute the succession and invade. As a result the Chitral garrison had to be stiffened, and the fort occupied by British forces. This stronghold was reputed to be the most impregnable fortress in the entire region of Dardistan. But possibly more important, it was a cultural and ancestral home for the tribes of Chitral. Occupation by the British or for that matter anyone else other than the Mehtar was an indignity the Chitralis could not ignore; compared to this outrage, genocide was an acceptable regular occurrence. There followed mass desertions of men enrolled to the British cause, and when Sher Afzul appeared suddenly from exile in Afghanistan, the scene was set for possibly the most famous siege of all time.

Robertson, the acting Governor of Gilgit, now ensconced with 550 persons in Chitral fort, had monumental problems to solve during the seven-week siege. On 3 March, a reconnaissance attempt from the fort led to the death of one British officer, two Kashmiri officers and twenty-three other ranks, whilst thirty men were seriously wounded. In early April attackers set fire to one side of the fort, and in mid-April a tunnel was discovered under the walls, just completed and ready for use.

When news of the siege was smuggled out, frantic consultations were held. The increasing severity of deprivations can be imagined, but it was estimated that no relief could be achieved until the end of April. It was fervently hoped that the high passes and winter snows would not add to the already formidable difficulties; the melting snow alone would cause severe enough problems. The final unknown in the equation was the resistance that would be mounted against the relief columns; it was impossible to predict its strength, but obviously it would be persistent, continual and determined. Relief columns from Kashmir via Bunji and Gilgit, or via Babusar, Chilas and Gilgit would take too long. Of two possibilities, the first was to take the route over the Malakand and Lowarai passes surveyed twelve years previously by MacNair. The second and almost inconceivable alternative was to permit a previously undistinguished and now elderly Colonel James Kelly, then resident in Chilas, to mount a subsidiary attempt over the Shandur Pass with whatever troops and supplies he could muster and descend on Chitral from the north. It was decided that both routes should be attempted, but the latter alternative was given little or no chance of success.

Kelly's march is now known to be one of the greatest triumphs in British military history. Even when compared with tales of Gordon of Khartoum, this relief expedition was two orders of magnitude greater in terms of endurance and problems of execution. Even expeditioners who frequent the valleys and mountains in mid-summer cannot fully appreciate the difficulties that Kelly had to overcome.

Kelly set out from Gilgit on March 23 with 100 Hunza-Nagar men, 400 men of a road-building regiment and a Kashmir mountain battery consisting of two guns. Soon after reaching the deep, melting snows, their troubles began and they had to retreat temporarily and re-assemble, many of their porters having deserted. It was unheard-of to cross the Shandur Pass at such a time of the year. Several alternative methods for transporting the two mountain guns were tried and all failed, but the enthusiasm of the battery commander, Lieutenant Cosmo Stewart, now acted as a catalyst. The guns were stripped, and although the barrels weighed 250kg (a quarter of a ton), the men in teams of four staggered along in 50m (164ft) relays.

It took three days to cross the pass through sometimes shoulder-deep snow, but all the difficulties of altitude, cold, frustration, lack of adequate nourishment and snow-blindness were overcome. The resistance on the way to Mastuj below the pass was insignificant in comparison with the difficulties of the climb up but a major battle took place a few kilometres below Mastuj at Nisa Gol, a defensive position which was considered impregnable. The topography was, indeed, even more horrendous than at Nilt, but the lesson had been learned. Kelly employed careful reconnaissance, using the various experts in the group and the skilled mountaineers of the Hunza-Nagar forces to the full, and the outcome was a rapid victory. Kelly showed his abilities by refusing to take risks or expose his men unnecessarily, adopting an uncomplicated strategy and coupling this with a humane execution of the task in hand. Thanks to his skill, few men were exposed to danger, and fewer still were killed in battle.

Kelly's forces arrived triumphant in Chitral on 20 April. Two days previously Sher Afzul had fled, realising all was lost. The farthest corner of the British Empire in India had at last been reached. The borders between India (which now included Chitral), Afghanistan, China, Russia and Kashmir (which encompassed Dardistan) were now generally agreed and would remain for the next fifty years, until the partition of India.

The bitter disputes that followed the British withdrawal from India left their mark even to this day. The Maharaja of Kashmir, a Hindu, was ruler of a Mohammedan people, who, given a free choice, might well have opted to join the Islamic state of Pakistan. But the Maharaja, seeking to obtain the political arrangement most advantageous for himself and his family, threw in his lot with India. The Dards rebelled and advanced on Gilgit, where they dispossessed the Maharaja's representatives before marching on towards Srinagar. The UN Security Council called for a cease-fire, to be followed by a plebiscite; the former was agreed, but the latter has still to take place. One product of the dispute is that the Northern Areas, which include Baltistan (with its capital Skardu), Hunza, Gilgit, Chilas, Yasin and Ishkuman, are still not fully incorporated into the provincial system of Pakistan, whilst the issue of Kashmir also remains unsettled. More serious from the Pakistani point of view, is that the Indians now control the headwaters of the Indus, the life-blood of Pakistan. Large-scale irrigation schemes in Ladakh and Kashmir could have a serious effect on the Pakistan economy, and although a judgement of the International Court redressed the balance to some extent, Pakistan still feels that she is left in an unenviable position.

The history of the area certainly had strong influences on the Karakoram Project. The Indian press viewed us with suspicion, and the Russians denounced the expedition as CIA-inspired. The Chinese, whilst giving us permission to travel inside China, did not wish to aggravate the delicate situation which had developed in Afghanistan just before our arrival (though long after the initiation of our plans), which had led to the presence of numerous refugees in Baluchistan and the North-West Frontier Province and Northern Areas of Pakistan. It is to the credit of Pakistan that we were permitted to proceed with our scientific studies, though permission was not to be achieved without difficulty. Certainly our work was closely monitored; I should not be at all surprised if accounts of our adventures were to be found on the foreign affairs desks of several countries concerned about developments in an area which, in my opinion, ought to become an internationally accepted national park. It is interesting to reflect that this was also the dream of Leitner, who unsuccessfully battled for the Dards to be left to their own devices, from the moment he crossed the Indus at Bunji until his death.

# Appendix II: Organisation and Membership

## International Project Committee
Dr John B. Auden (Chairman)
Dr G. Colin L. Bertram
Professor Eric H. Brown
Sir Douglas Busk
Dr Andrew Goudie (Deputy Leader)
George Greenfield
Lt. Col. David Hall
Brigadier George Hardy
John Hemming (Director RGS)
Lord Hunt of Llanfair Waterdine (President RGS)
Professor Keith J. Miller (Leader)
Mr Michael Ward
Nigel de N. Winser (Deputy Leader and Secretary)

## Consultants
*Geomorphological Research*
Professor E. H. Brown
Professor K. J. Gregory
Professor N. J. Stevens
Dr R. J. Price
Dr J. M. Grove
Dr D. Q. Bowen

*Housing and Natural Hazards Research*
Professor N. Ambraseys
Professor W. G. V. Balchin

Professor O. Koenigsberger
D. J. Dowrick

*Radar Ice-Depth Sounding Research*
Sir Vivian Fuchs
Dr S. Evans
Dr D. Drewry
Dr G. de Q. Robin
Dr C. W. M. Swithinbank
Dr C. M. Doake

*Geological Research*
Asrarullah (Geological Survey of Pakistan)
Dr B. F. Windley

*Survey Research*
P. G. Mott
J. G. Olliver
A. L. Allen
Mianmohammed Sharif (Survey of Pakistan)

## Pakistan Liaison
David Latter, British Council Representative, Islamabad

## Distinguished Visitors
H. E. The British Ambassador Oliver Forster
Major General Mushtaq Ahmad Gill
Lord and Lady Hunt
Lord Shackleton

## Members of Expedition

| | Profession/Team | Age |
|---|---|---|
| Abbas, S. Ghazanfar | Deputy Director<br>Geological Survey of Pakistan<br>Geology Team | 34 |
| Ahmad, Shabbir | Lecturer<br>Quaid-i-Azam University<br>Seismology Team | 25 |
| Akbar, Khurshid | Assistant Geophysicist<br>Geological Survey of Pakistan<br>Seismology Team | 27 |
| Allen, John | Postgraduate research student<br>Sheffield University<br>Survey Team | 22 |
| Aman, Ashraf | Mountaineer<br>Rawalpindi Tourist Industries<br>Survey Team | 35 |
| Atkinson, Nigel | Surveyor<br>Hunting Surveys and Consultants Ltd<br>Survey Team | 26 |

| | | |
|---|---|---|
| Awan, Abdul Razzaq | Surveyor<br>Survey of Pakistan<br>Survey Team | 32 |
| Bilham, Roger | Geophysicist<br>Lamont-Doherty Geological Observatory<br>Survey Team | 35 |
| Bishop, James | Civil Engineer<br>Sir Alexander Gibb and Partners<br>Survey Team | 30 |
| Brunsden, Denys | University Reader<br>King's College, University of London<br>Geomorphology Team | 44 |
| Charlesworth, Ronald | Freelance Photographer<br>Three Arrows Limited<br>Film Unit | 48 |
| Chen, Jianming | Engineer and Surveyor<br>Institute of Glaciology and Cryopedology,<br>    Lanzhou<br>Survey Team | 43 |
| Collins, David Nigel | University Lecturer<br>Manchester University<br>Geomorphology Team | 32 |
| Colvill, Alan John | Chartered Accountant/Land Surveyor<br>University of Colorado<br>Survey Team | 39 |
| Crompton, Thomas Oliver | University Lecturer<br>University College, University of London<br>Survey Team | 33 |
| Davis, Ian Robert | Principal Lecturer<br>Oxford Polytechnic<br>Director, Housing and Natural Hazards Team | 43 |
| Davison, Ian | Geologist<br>British National Oil Corporation<br>Seismology Team | 25 |
| Derbyshire, Edward | University Reader<br>University of Keele<br>Geomorphology Team | 48 |
| Dong, Zhi Bin | University Lecturer<br>Lanzhou University<br>Radar Ice-Depth Sounding Team | 43 |
| Durrani, Nasir Ali | Professor of Geology<br>University of Baluchistan, Quetta<br>Geology Team | 41 |
| Farooq, Mohamed | Surveyor<br>Survey of Pakistan<br>Survey Team | 36 |
| Ferguson, Robert Ian | University Senior Lecturer<br>Stirling University<br>Geomorphology Team | 34 |
| Ferrari, Ronald Leslie | University Lecturer<br>Cambridge University<br>Radar Ice-Depth Sounding Team | 50 |
| Francis, Marcus | Research Engineer<br>British Hydraulics Research Association, Cranfield<br>Co-Director, Radar Ice-Depth Sounding Team | 30 |

| | | |
|---|---|---|
| Ghauri, Arif Ali Khan | Associate Professor<br>Peshawar University<br>Geology Team | 42 |
| Giles, David Peter Vaughan Lindsey | Doctor<br>Medical Practice, Bude<br>Director, Medical Team | 45 |
| Goudie, Andrew Shaw | University Lecturer<br>Oxford University<br>Deputy Leader and Director, Geomorphology<br>    Team | 35 |
| Holmes, Robert Edward | Photographer (Stills)<br>Freelance Photographer<br>Film Unit | 37 |
| Hughes, Richard Edward | Geotechnical Archaeologist<br>Ove Arup & Partners<br>Housing and Natural Hazards Team | 31 |
| Illi, Dieter | University Lecturer<br>Zurich University<br>Housing and Natural Hazards Team | 38 |
| Islam, Shaukat | University Lecturer<br>Peshawar University<br>Botanist | 40 |
| Israr-ud-Din | Associate Professor<br>Peshawar University<br>Housing and Natural Hazards Team | 42 |
| Jan, Qasim M. | Associate Professor<br>Peshawar University<br>Geology Team | 37 |
| Jackson, James Anthony | Research Fellow<br>Queens' College, Cambridge University<br>Joint Director, Seismology Team | 25 |
| Jones, David Keith Crozier | University Lecturer<br>London School of Economics, University of London<br>Geomorphology Team | 40 |
| Khan, Islam M. | University Lecturer<br>Peshawar University<br>Botanist | 46 |
| Khan, Muhammad Zakir | Major, Pakistan Army<br>Liaison Officer | 32 |
| Khattak, Rehman | Electrical Engineer<br>Pakistan Atomic Energy Commission<br>Seismology Team | 40 |
| King, Geoffrey Charles Plume | Senior Research Assistant<br>Cambridge University<br>Joint Director, Seismology Team | 36 |
| Li, Jijun | Associate Professor<br>Lanzhou University<br>Geomorphology Team | 47 |
| Lin, Ban Zuo | Engineer<br>Institute of Geophysics, Beijing<br>Seismology Team and<br>    Co-leader of Chinese Team | 44 |
| Massil, Helen | Doctor<br>Medical Practitioner<br>Medical Team | 25 |

| | | |
|---|---|---|
| Miller, Keith John | Professor of Mechanical Engineering<br>Sheffield University<br>Leader of Expedition | 48 |
| Moughtin, James Clifford | Professor of Planning<br>Nottingham University<br>Housing and Natural Hazards Team | 48 |
| Moughtin, Timothy | Assistant<br>Housing and Natural Hazards Team | 16 |
| Muir Wood, Robert | Journalist/Geologist<br>Writer for *New Scientist* | 28 |
| Musil, George Jiři | Postgraduate Research Student<br>British Antarctic Survey<br>Radar Ice-Depth Sounding Team | 22 |
| Nash, David Francis Tyris | University Lecturer<br>Bristol University<br>Housing and Natural Hazards Team<br>(Mrs Nash also gave valuable service to the Project) | 31 |
| Nunn, Paul James | Mountaineer/Lecturer<br>Sheffield Polytechnic<br>Film Unit/Radar Ice-Depth Sounding Team | 37 |
| Oswald, Gordon Kenneth Andrew | Electronics Research Engineer<br>Cambridge Consultants Ltd<br>Director, Radar Ice-Depth Sounding Team | 31 |
| Perrott, Frances Alayne | Lecturer<br>Oxford University<br>Geomorphology Team | 29 |
| Rajab, Ali | Sirdar of Porters<br>Survey Team | 32 |
| Rana, Javaid Akhtar | Major, Pakistan Army<br>Liaison Officer | 31 |
| Redhead, Charles Stephen | Driver<br>Transport Team | 23 |
| Rehman, Mohammad Abdul | Geologist<br>Pakistan Atomic Energy Commission<br>Geology Team | 39 |
| Rendell, Helen | Lecturer<br>Sussex University<br>Geomorphology Team | 29 |
| Riley, Anthony | Freelance Cameraman<br>Three Arrows Limited<br>Film Unit | 37 |
| Said, Mohammed | Professor of Geography<br>Peshawar University<br>Geomorphology Team | 44 |
| Smith, Edward Whittaker | Lecturer<br>University of Manchester<br>   Institute of Science and Technology<br>Survey Team | 35 |
| D' Souza, Frances | Director<br>International Disaster Institute<br>Housing and Natural Hazards Team | 38 |

| | | |
|---|---|---|
| Spence, Robin John | Lecturer<br>Cambridge University<br>Housing and Natural Hazards Team<br>(Dr Spence had his research student, Andrew<br>Coburn, assist for a short period) | 39 |
| Stoodley, Robert Arthur | Chairman and Managing Director<br>Manor National Group Motors Ltd<br>Transport Team Manager | 56 |
| Tahir, Iqbal | Major, Pakistan Army<br>Liaison Officer | 29 |
| Tahirkheli, Rashid Ahmad Khan | Professor of Geology<br>Peshawar University<br>Director, Geology Team and<br>    Leader of Pakistani Scientists | 53 |
| Walton, Jonathan Lancelot William | Land Surveyor<br>University College, London University<br>Director, Survey Team | 30 |
| Waters, Ronald Sidney | Professor of Geography<br>Sheffield University<br>Geomorphology Team | 58 |
| Wesley-Smith, Shane | Administrator<br>Royal Geographical Society<br>Administrative Team | 24 |
| Whalley, Brian | Lecturer<br>Queens University, Belfast<br>Geomorphology Team | 33 |
| Winser, Nigel de Northop | Administrator<br>Royal Geographical Society<br>Deputy Leader and Director, Administrative Team | 28 |
| Xu, Shuying | Associate Professor<br>Lanzhou University<br>Geomorphology Team | 46 |
| Yielding, Graham | Postgraduate research student<br>Cambridge University<br>Seismology Team | 22 |
| Zafar, Hashmat | Hydrogeologist<br>Water and Power Development Authority<br>Geomorphology Team | 31 |
| Zhang, Xiangsong | Associate Professor<br>Institute of Glaciology and Cryopedology,<br>    Lanzhou<br>Radar Ice-Depth Sounding Team and Co-Leader of<br>Chinese Team | 44 |

# Appendix III: Sponsors

*Sponsors*
Government of Pakistan
Academia Sinica, Beijing
Royal Geographical Society

*Financial Support*
Barclays Bank International
Binnie & Partners
British Council
William A Cadbury Charitable Trust
Carnegie Trust for the Universities of Scotland
Winston Churchill Memorial Trust
The Drapers' Company
The Goldsmiths' Company
Harrods Ltd
Longdin & Browning
Manchester Geographical Society
Mount Everest Foundation
National Geographic Society
Natural Environment Research Council
Overseas Development Administration
Albert Reckitt Charitable Trust
Rio Tinto Zinc Corporation
Royal Geographical Society
Royal Institution of Chartered Surveyors
Royal Scottish Geographical Society
The Royal Society
Science Research Council
Wexas International
George Wimpey & Co. Ltd
United Nations

*Medical Supplies*
Abbott Laboratories Ltd
A D International Ltd
Allen & Hanburys Ltd
Astra Chemicals Ltd
Bayer (UK) Ltd
Beecham Research Laboratories
Bencard
Boehringer Ingleheim Ltd
Boots Company Ltd
Calmic Medical Division of the Wellcome
  Foundation
Ciba Laboratories
Davis & Geck
Dista Products
Dome Laboratories
Downs Surgical Ltd
Duncan Flockhart & Co Ltd
Eli Lily & Co Ltd
Fair Laboratories
Farley Health Products

Geigy Pharmaceuticals
Glaxo Laboratories
Hoechst Pharmaceuticals
Imperial Chemical Industries
Janseen Pharmaceuticals Ltd
Johnson & Johnson Ltd
Kirby Warwick Ltd
Laboratories for Applied Biology
Lederle Laboratories
Leo Laboratories
May & Baker Ltd
Merck Sharp & Dohme Ltd
Miles Laboratories
Montedison Pharmaceuticals Ltd
Nicholas Laboratories
Novo Laboratories
Lakeland Plastics
Parke Davis & Co Ltd
Pfizer Ltd
Pharmacia (GB) Ltd
Pharmax Ltd
A H Robins Co Ltd
Roche Products Ltd
Roussel
Reckitt & Colman Pharmaceuticals
Scholl (UK) Ltd
Serle Laboratories
E R Squibb & Sons Ltd
Smith & Nephew Pharmaceuticals
Smith Kline & French Laboratories
Strentex Fabrics Ltd
Chas. Thackery Ltd
3M UK Ltd
Upjohn Ltd
Vernon Carus Ltd
Wander Pharmaceuticals Ltd
Wlm Warner Ltd
Warner-Lambert Ltd
WB Pharmaceuticals
The Wellcome Foundation
Winthrop Laboratories
Wyeth Laboratories

*Foodstuffs*
British Food Export Council
Coca Cola Corporation
Colmans Foods
Drinkmaster Ltd
Frank Cooper
H J Heinz
The Honey Bureau & Gales Honey
John West Foods
Nabisco Ltd
Raven Foods

Ryvita
Sabatani & Taylor Associated for Schwartz Spices
St Ivel Ltd
Tate & Lyle
Weetabix Ltd
Whitworths Holdings
Unilever Export Ltd

## Expedition Equipment

Berec International
Brillo Manufacturing Co
Briton Chadwick Ltd
Bryant & May
Burgess Rucksacs
Camping Gaz (GB) Ltd
Damart Thermal Underwear
Estercol Sales Ltd
Flatetec Ltd
Hamish Hamilton
Jen Shoes & Boots
Karrimor International
Kitbin
Laughton & Sons
Lacrinoid Ltd
Racal Radios
Mountain Equipment
Mitsui Machinery Sales
Northcape Textiles
Nevisport
Nikwax
A B Optimus Ltd
Penguin Books
Phillips of Axminster
Portacel Waterfilters
Pindisports
Rohan
Spacecoat Garments
Stanley Tools Ltd
Supreme Plastics
Swains Packaging
Thor Hammer Co
Thermos Ltd
Tog 24
Tri Wall Containers
Troll Safety Equipment
Vango Scotland Ltd

## Transport

Anchor Line Ship Management Ltd
Associated Tyre Specialists
Automobile Association (Stockport)

British Airways
Brown Jenkinson (Liverpool) Ltd
Burmah Castrol Ltd
Caltex Oil (Pakistan) Ltd
Fairey Engineering Ltd
Land Rover Ltd
Lifting Gear Hire Ltd
Lloyds Industries
Manor National Group Motors
Michelin Tyre Company Ltd
Pakistan National Shipping Corp
Pakistan State Oil Co Ltd
Protofram Ltd
Trailvan Ltd
Unipart Ltd

## Insurance Services

Sedgewick International

## Photographic Work

Classic Vases
C & J Clark Ltd
Geographical Magazine
Japanese Cameras (Minolta)
Kodak Ltd
Larson Enterprises
Laptech Studies
National Geographic
Quest Vest
Three Arrows Films
Tenba Bags

## Survey Equipment

Aga Geotronics
British Aluminium
Casio Electronics
Decca Surveys
Hunting Surveys
Mapping & Charting Establishment RE
School of Military Survey
SLD Sitelink Services
Survey & General Instruments
Tellurometer (UK) Ltd
Wild Heerbrugg

## Other Scientific Equipment

Avon Rubber Company
Chloride Batteries (Pakistan) Ltd
Hewlett Packard Ltd (Geneva)
Perex Ltd

# Appendix IV: Scientific Publications Related to the IKP 1980

Proceedings of the International Karakoram Project, to be published in 1982/1983.

Vol I: Proceedings of the Conference held at the Quaid-i-Azam University, Islamabad, June 1980.

1 Inauguration address  *The President of Pakistan, General Mohammad Zia-ul-Haq*
2 Address on behalf of the Royal Geographical Society  *Lord Shackleton KG*
3 Address on behalf of the Government of Pakistan  *The Minister of Education, Mr Muhammad Ali Khan*
4 Some recent technological advances applied to problems in earth sciences  *K. J. Miller*
5 Point-positioning by doppler satellite  *J. P. Allen*
6 Survey and analysis systems for the Vatnajökull ice-depth sounding expedition 1977  *J. F. Bishop, K. J. Miller*
7 The Pakistan (India) to Russia triangulation connexion: past and projected error analysis  *T. O. Crompton*
8 Special techniques for surveying on moving terrain  *J. Walton*
9 Variation of the Batura glacier's surface by repeated surveying and mapping  *Chen Jianming*
10 Indo-Asian convergence and the 1913 survey line connecting the Indian and Russian triangulation surveys  *R. Bilham, D. Simpson*
11 Tectonic studies in the Alpine Himalayan Belt  *G. King, J. Jackson*
12 Basement fault reactivation in young fold mountain belts  *J. Jackson*
13 Earthquake activity, fault plane solution and plate tectonics of the Himalaya and surrounding regions  *Teng Ji Wen and Lin Ban Zuo*
14 The mechanics of fracture applied to ice  *K. J. Miller*
15 Fracture toughness of glacier ice  *R. M. Andrews, A. R. McGregor, K. J. Miller*
16 Electronic design and performance of an impulse radar ice-depth sounding system used on the Vatnajökull ice-cap, Iceland  *A. D. Cumming, R. L. Ferrari, G. Owen*
17 Results of impulse radar ice-depth sounding on the Vatnajökull ice-cap, Iceland  *J. F. Bishop, R. L. Ferrari, K. J. Miller*
18 Recent variations of some glaciers in the Karakoram mountains  *Zhang Xiangsong*
19 Some studies of the Batura glacier  *Shi Yafeng, Zhang Xiangsong*
20 A surging advance of Balt Bare glacier in the Karakoram mountains  *Wang Wenying, Huang Maochuan, Chen Jianming*
21 Some observations on glacier surges with notes on an example in East Greenland  *Alan Colvill*
22 The distribution of glaciers on the Qinhai-Xizang Plateau and its relationship with atmospheric circulation  *Li Jijun and Xu Shuying*
23 The geomorphology of high magnitude – low frequency events in the Karakoram mountains  *D. Brunsden, D. K. C. Jones*

17   Pleistocene glacial history of the Hunza Valley   *E. Derbyshire et al.*

18   Sediment yields of the Hunza Valley   *R. I. Ferguson*

19   Salt efflorescences and salt weathering in the Hunza Valley   *A. S. Goudie*

20   Hydrology and hydrochemistry of glacial meltwaters   *D. N. Collins*

21   Rock temperature observations and chemical weathering in the Hunza region   *W. B. Whalley, J. P. McGreevy, R. I. Ferguson*

22   Hazards of the Karakoram Highway   *D. Brunsden, A. S. Goudie, D. K. C. Jones*

23   Lake deposits of Borritgil   *F. A. Perrott*

24   Impulse radar ice-depth sounding on the Hispar Glacier   *Z. B. Dong, R. L. Ferrari, M. R. Francis, G. Musil, G. K. A. Oswald, X. Zhang*

25   Ice-depth radio echo-sounding techniques employed on the Hispar and Ghulkin glaciers   *G. Oswald*

26   Impulse radar ice-depth sounding of the Ghulkin glacier   *M. R. Francis, K. J. Miller, Z. B. Dong*

27   On the present glaciers in the Karakoram   *X. Zhang, D. Mi*

28   The survey work of the International Karakoram Project 1980   *J. Chen, T. O. Crompton, J. Walton, R. Bilham*

29   The calculation of crustal deformation and strain from classical survey measurements   *T. O. Crompton*

30   A preliminary study of ancient trees in the Hunza Valley and their dendroclimatic potential   *R. Bilham, G. B. Pant, G. C. Jacoby*

31   A microearthquake survey in the Karakoram   *G. Yielding, S. Ahmad, I. Davison, J. A. Jackson, R. Khattack, A. Khurshid, G. C. P. King, B. Z. Lin*

32   Source studies of the 1972 Gupis and 1974 Patan earthquakes   *J. A. Jackson*

33   Studies of large earthquakes in the Alpine-Himalayan mountain belt   *G. C. P. King*

34   A summary of the geological research undertaken by the IKP 1980   *R. A. K. Tahirkheli et al*

35   A new look at the geology of the Karakoram   *R. A. K. Tahirkheli*

36   The geology of the south-central Karakoram and Kohistan   *M. Coward, Q. M. Jan, B. F. Windley*

37   A science of mountains   *R. Muir Wood*

38   A preliminary discussion of the landform development and quaternary glaciation from Potwar to the Hunza Valley   *J. Li, S. Xu*

39   Glacial and pro-glacial deposits and their evolution in the Hunza Valley   *S. Xu, J. Li*

40   A medical survey taken during the IKP 1980   *D. Giles, H. Massil*

41   Stress field plate motion of the Karakoram and surrounding regions   *P. Y. Shu, B. Z. Lin*

Articles published in *Geographical Magazine*

'Fearful Landscape of the Karakoram', *Andrew Goudie, vol. 53 (No. 5), pp. 306–12*

'High Road to Lonely Shimshal', *Robert Muir Wood and John Hunt, vol. 53 (No. 8), pp. 504–10*

'Highway beneath the Ghulkin', *Edward Derbyshire and Keith Miller, vol. 53 (No. 11), pp. 626–35*

CONTINENTS IN COLLISION

Articles by Dr Robert Muir Wood published in *New Scientist*

'Science goes to the Karakorum', *vol. 88 (1226), pp. 374–7*
'Islands at the top of the mountains', *vol. 89 (1238), pp. 274–7*
'The moving mountains', *vol. 89 (1242), pp. 540–3*
'Decay in the Karakorum', *vol. 89 (1246), pp. 820–3*
'Hard times in the mountains', *vol. 90 (1253), pp. 414–17*
'Highway to the top of the world', *vol. 91 (1270), pp. 656–8*

# Appendix V: History of the Ascent of Rakaposhi

No article, book or lecture on Hunza or Gilgit would be complete without including some detail on Rakaposhi (7,788m, 25,550ft), the peak that dominates the area. This short account of attempts to climb the mountain is by Michal Vyvyan, one of the first to tackle Rakaposhi.

'The majority of the attempts to climb Rakaposhi have been made via the ridges on the south and west (see Fig. below) rather than via the daunting north face. The first recorded attempt was by Martin Conway in 1892. He visited the southern approaches during a superbly recorded reconnaissance of the Karakoram and correctly identified the south-west ridge as offering the best route to the summit. Rakaposhi was not reconnoitred again from a mountaineering point of view until Secord and Vyvyan went to the Dara Khush (Happy Valley) below the Biro Glacier in 1938 and climbed Jaglot Peak (the north-west peak) via its western spur. They also identified the most hopeful route to the summit as lying along the south-west ridge and named a principal feature of this ridge the Monk's Head, on account of its cowl-like appearance.

In 1947 Secord went back to Rakaposhi with Tilman, Gyr, and Kappeler, the last two of whom had explored other approaches to the peak in 1946. They began with an attempt on the south-west ridge via the Monk's Head. Unpromising weather deterred the party at about 6,100m (20,000ft) so Tilman and Gyr returned to the Biro Glacier and climbed up to a point on the north-west ridge which connects the Jaglot Peak to the summit complex. But the route was not pursued because of the ever-present danger of avalanches on the approaches to the ridge. In 1954 and 1956 two more unsuccessful attempts were made via the south-west branch – or spur – of the south ridge.

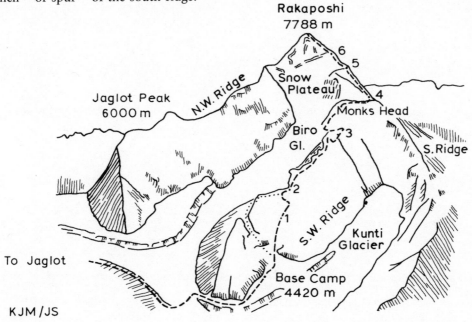

KJM/JS

*Rakaposhi*

In 1954 Tissières, Band and Fisher, after reclimbing Jaglot Peak, rejoined the rest of their party on the south-west ridge and climbed it as far as the edge of the Monk's Head – but the weather then forbade further reconnaissance. Two years later Banks's powerful British-American quartet passed this landmark but were frustrated by avalanche conditions at an estimated 7,160m (23,500ft). Then, on 25 June 1958, Banks himself, with Patey, was finally successful in reaching the summit from a sixth camp on this route supported by a party of eight which faced 'terrible conditions of snow and drift' (see numbered route on figure). Both Banks's teams included active Pakistani climbers and official co-operation may have slightly compensated for the severe weather by easing logistical problems.

In 1964 Lynam's Irish party made some progress along the north-west ridge reaching it, like Tilman, from the Biro Glacier. This route was eventually completed in 1979 by a Polish–Pakistani party led by Kowalewski and Sher Khan. These two, with Pietrowski, achieved the second ascent of the mountain on 1 July followed by three others on 2 July. Then on 5 July the summit was reached again by two Polish girls, Czerwinska and Palmowska, climbing individually and leaving the men, it seems, at Camp 3.

Almost simultaneously with the Polish success a Japanese expedition from Waseda University, led by A. Ohtani, put a party on the summit on 2 August (thus making the third ascent of the peak). They followed a buttress on the daunting North side of the mountain which had been attempted by Herrligkoffer's German parties in 1971 and 1973. Details of this fine Japanese achievement are not yet available.'

# References

**BEGINNINGS**

1 'Under-Ice Volcanoes', *The Geographical Journal*, 1979, Vol. 145, 36–55.
2 'Traverse of the Staunings Alps', *Alpine Journal*, 1976, 143–53.

**JOURNEY TO SHIMSHAL**

1 See E. F. Knight, *Where Three Empires Meet*, Longmans.

**THE GHULKIN GLACIER**

1 *Simple Surveying Techniques for Small Expeditions*, one of a number of pamphlets published by the Royal Geographical Society.

**APPENDIX I HISTORY OF EXPLORATION**

1 *The Gilgit Game* and *When Men and Mountains Meet* by J. Keay (John Murray) are recommended reading.
2 See Eric Newby, *A Short Walk in the Hindu Kush*, Secker and Warburg.

**FURTHER READING**

'Recherches géologiques dans les Chaines alpines de l'Asie du sud-ouest', *Mem. hors-serie N°
8 de la Société géologique de France*, 1977.
*Geodynamics of Pakistan*, Ed Abdul Farah and Kees A. DeJong, Geological Survey of Pakistan, 1979.
*Blank on the Map*, Eric Shipton, Hodder & Stoughton, 1938.
*High Mountains and Cold Seas: A Biography of H. W. Tilman*, J. R. L. Anderson, Victor Gollancz, 1980.

# Index

References in italic refer to figures in the text

# INDEX